On Fuzzy Cluster Relevance

On Addressing Cluster Count Ambiguity using Fuzzy Cluster Relevance Factors

ARASH ABADPOUR

Arash Abadpour (arash@abadpour.com) is with Intellijoint Surgical Inc., Waterloo, Canada.

The author acknowledges that this text has been submitted for publication and that this file will be deleted upon publication.

Contents

1 **Introduction**   1

2 **Literature review**   5
    2.1 Notion of Membership 5
    2.2 Prototype-based Clustering 7
    2.3 Robustification 9
    2.4 Number of Clusters and Cluster Fuzziness 11
    2.5 Weighted Clustering 11

3 **Developed Method**   13
    3.1 Design Methodology 13
    3.2 Model Preliminaries 14
    3.3 Cluster Fuzziness 17
    3.4 Assessment of Loss 19
    3.5 Solution Strategy 22
    3.6 Outlier Detection and Classification 24
    3.7 Cluster Maintenance 24
    3.8 Determination of U and λ 24
    3.9 Implementation Notes 26

4 **Experimental Results**   31
    4.1 Algorithm Overview 31
    4.2 φ Diagnostics 33
    4.3 Impact of κ 36
    4.4 Comparative Assessment 37

5 Conclusions		**47**
A Problem Classes		**51**
A.1	Circle-Finding in 2D Data (2dc)	55
A.2	Euclidean Clustering of 2D Data (2de)	55
A.3	Line-Finding in 2D Data (2dl)	56
A.4	Plane-Finding in Range Data (3dpp)	56
A.5	Segmentation of a Color Image (ics)	57
A.6	Histogram-based Segmentation of a Grayscale Image (ighe)	57

List of Figures

3.1 Pictorial representation of the concepts of *Problem Class* and *Problem Instance* in the context of the present work and their relationship with the method developed in this paper. 14

3.2 The process of selecting φ_1, based on which the value of λ is determined. 25

3.3 The process of selecting φ_0, based on which the value of U is determined. 25

3.4 Flow of the variables in the Alternating Optimization process developed in this paper. 27

3.5 Implementation of the flow exhibited in Figure 3.4. 28

4.1 Overview of the developed algorithm from an input/output perspective for a 2dl problem instance. (a) Input set of data items. (b) Converged clusters. (c) Classification of input data items into clusters and outliers. (d), (e), (f), and (g) Internal state of the clustering algorithm. (d) Histogram of f_{nc} during the convergence of the algorithm. (e) Histogram of p_n during the convergence of the algorithm. (f) Histogram of r_c during the convergence of the algorithm. (g) Value of $C_\circ$ compared to the values of C and $E\{\mathbf{C}\}$ during the convergence of the algorithm. 32

4.2 Assessment of the impact of the value of φ on the results of the developed algorithm for the 2de problem class. (a) $\varphi = \frac{1}{10}\varphi_{2de}$. (b) $\varphi = \varphi_{2de}$. (c) $\varphi = 10\varphi_{2de}$. Here, φ_{2de} denotes the optimal value of φ for the 2de problem class, as discussed in Appendix A.2. 34

4.3 Assessment of the impact of the value of φ on the results of the developed algorithm for the 2dl problem class. (a) $\varphi = \frac{1}{10}\varphi_{2dl}$. (b) $\varphi = \varphi_{2dl}$. (c) $\varphi = 10\varphi_{2dl}$. Here, φ_{2dl} denotes the optimal value of φ for the 2dl problem class, as discussed in Appendix A.3. 36

4.4 Summary of the execution results for a 2dc problem instance with three clusters and different κ values. (a) Number of clusters initialized at the beginning of the process. (b) Total elapsed time. 37

4.5 Initial and converged clusters for different κ values. Top) Clusters as they are initialized. Bottom) Converged clusters. 38
4.6 Pairs of results generated by the proposed method and FCM, carried side-by-side in order to allow comparison. (a) Input data items. (b) Output of the proposed method. (c) Output of FCM. 40
4.7 Pairs of results generated by the proposed method and FCM, carried side-by-side in order to allow comparison. (a) Input data items. (b) Output of the proposed method. (c) Output of FCM. The input data used in 3dpp experiments is courtesy of *Epson Edge, Epson Canada Limited.* 42
4.8 Pairs of results generated by the proposed method and FCM, carried side-by-side in order to allow comparison. (a) Input data items. (b) Output of the proposed method. (c) Output of FCM. 43
A.1 Geometrical models corresponding to the problem classes utilized in this paper. . . . 53

Abstract

Unsupervised clustering of a set of data items of an arbitrary model into clusters which comply with a particular model of homogeneity is a useful tool in many applications within many fields, including the general field of computer sciences. This problem, however, is in essence ill-posed, due to the fact that it concerns the determination of cluster representations as well as data item-to-cluster correspondences, two entities which are heavily entangled. The research community is well aware of this ambiguity and has responded with several important contributions, such as the fuzzy membership regime and varied approaches to robustification, among others. Some of these works, however, utilize models and parameters which are in essence based on the intuition of the researchers and result in the presence of regularization coefficients and variables and configuration parameters which are defined metaphorically. In this work, we utilize a derivation-based approach and perform Bayesian loss modeling in order to derive the objective function for a fuzzy clustering algorithm which utilizes fuzzy data. The third component of the fuzziness regime explored in this paper, which to the best of our knowledge is novel to this work, is the fuzzification of the clusters. We demonstrate that fuzzy clustering of fuzzy data into fuzzy clusters is a proper continuation of the movement within the data clustering community that resulted in important achievements in terms of the performance of clustering algorithms.

Chapter 1

Introduction

Unsupervised data clustering is the general problem of separating a set of data items into a number of homogenous subsets and potentially a set of outliers. When this concept is used in practice, however, several important questions are to be answered. For example, while unsupervised data clustering is utilized very frequently, it is generally executed for data items of different mathematical models and varied notions of homogeneity. Hence, it is significantly advantageous, from the perspective of algorithm and code re-usability, that we address data clustering in generic terms. In order to be able to achieve this generalization we need to abstract out data and cluster models and to discuss unsupervised data clustering at a mathematical application-indifferent level.

There has been a tendency in the research community to utilize metaphorically defined concepts and objective functions in order to propose solutions to the unsupervised data clustering problem. This practice in essence entails the composition of objective functions which *"makes sense"*. In fact, sometimes the heuristics behind the suggested loss models are not provided at all and the objective functions are given as is and justification is postponed to the analysis of experimental results. In this line of thought, the proposal of a novel clustering algorithm starts with stating an objective function and continues with providing a solution strategy, which generally utilizes an iterative process. Then, experimental results are carried and are used as *evidence* for the appropriateness of the heuristics utilized in the formalization of the problem.

We argue that while experimentation is an important component in the advancement of the field of data clustering, satisfying experimental results do not suffice. In other words, any unsupervised data clustering algorithm must be able to produce results which satisfy the expectations, but the fact that a satisfying set of experimental results exists does not provide evidence for the usability of the corresponding algorithm. This is, firstly, due to the fact that, practically, no algorithm can be examined against an inclusive set of problem classes. In fact, this field is so vast and uncharted that suggesting that a set of unsupervised data clustering problems is a representative of the field is a claim similar in size to the introduction of a novel algorithm. Additionally, from an epistemological perspective, the composition of a data clustering algorithm must follow a construction mechanism which denotes why a particular model is utilized and what the different components of it correspond to.

Arguably, a generic clustering algorithm involves three identities, i.e. the data items, the clusters, and the relationship between the data items and the clusters. In this context, a conventional hard-style clustering algorithm utilizes crisp models for these three identities. In other words, such an algorithm suggests that a non-fuzzy membership regime is a proper model for the relationship between the data items and the clusters, both of which are non-fuzzy. The extended class of fuzzy clustering algorithms, which were born out of FCM and its predecessors, however, add fuzziness to the relationship between the data items and the clusters and show that this transformation is greatly beneficial towards avoiding local minimums. In this framework, most data items neither fully belong to nor fully not belong to most clusters. They in fact belong to every cluster to some extent. Moreover, utilizing fuzzy data items is a relatively straightforward extension in the formulations. In this framework, fuzzy clustering algorithms are capable of clustering fuzzy data into non-fuzzy clusters with the stipulation that the relationship between the data items and the clusters is fuzzy.

The community has greatly benefited from establishing a fuzzy relationship between the data items and the clusters. The utilization of fuzzy data items, too, has allowed multiple classes of data items and clustering speed-up, in addition to addressing cases in which the data items are inherently fuzzy. The third pillar of this platform, i.e. fuzzy clusters, however, has, to the best of our knowledge, never been explored. Hence, the clusters have remained crisp objects which are either fully relevant or are not relevant at all. We argue that this unexploited opportunity can generate invaluable returns. In this line of thought, the clusters are to be transformed from objects with binary relevance status into entities with a fuzzy relevance model. We show in this paper that the fuzziness of the clusters evolves during the procedure and allows for more appropriate treatment of the inherent ambiguities in the context of unsupervised data clustering.

The rest of this paper is organized as follows. First, in Chapter 2, we review the related literature and then, in Chapter 3, we present the developed method. Subsequently, in Chapter 4, we provide experimental results produced by the developed method and, in Chapter 5, we carry the concluding remarks.

Chapter 2

Literature review

2.1 Notion of Membership

The notion of membership is a key point of distinction between different clustering frameworks. Essentially, membership may be *Hard* or *Fuzzy*. Within the context of hard membership, each data item belongs to one cluster and it is different from all other clusters. The fuzzy membership regime, however, maintains that each data item in fact belongs to all clusters, with the stipulation that the degree of membership to different clusters is different. K-means [1] and Hard C-means (HCM) [2] clustering algorithms, for example, utilize hard membership values. The reader is referred to [3] and the references therein for a history of K-means clustering and other methods closely related to it. Iterative Self-Organizing Data Clustering (ISODATA) [4] is a hard clustering algorithm as well.

With the introduction of Fuzzy Theory [5], many researchers incorporated this more natural notion into clustering algorithms [6, 7, 8]. The premise for employing a fuzzy clustering algorithm is that fuzzy membership is more applicable in practical settings, where, generally, no distinct line of separation is present between the clusters. Additionally, from a practical perspective, it is observed that hard clustering techniques are extremely more prone to falling into local minima [9]. The reader is referred to [10, 11] for the wide array of fuzzy clustering methods developed in the past few decades.

Initial work on fuzzy clustering was done by Ruspini [12] and Dunn [13] and it was then generalized by Bezdek [10] into Fuzzy C-means (FCM). In FCM, data items, which are denoted as $x_1, \cdots, x_N$, belong to $\mathbb{R}^k$ and clusters, which are identified as $\psi_1, \cdots, \psi_C$, are represented as points in $\mathbb{R}^k$. FCM makes the assumption that the number of clusters, C, is known through a separate process or expert opinion and minimizes the following objective function,

$$\Delta = \sum_{c=1}^{C} \sum_{n=1}^{N} f_{nc}^m \|x_n - \psi_c\|^2. \tag{2.1}$$

This objective function is heuristically suggested to result in appropriate clustering results and is constrained by,

$$\sum_{c=1}^{C} f_{nc} = 1, \forall n. \tag{2.2}$$

Here, $f_{nc} \in [0,1]$ denotes the membership of x_n to ψ_c.

We note that the fact that membership values are binary in hard clustering approaches is not just because these approaches impose the $f_{nc} \in \{0,1\}$ constraint. In fact, merely relaxing this constraint into $f_{nc} \in [0,1]$ does not lead to a fuzzy assignment regime between the data items and the clusters [14]. Nevertheless, Borgelt [14] reviews the possibility of achieving fuzzy membership values through utilizing a regularization term which penalizes crisp or close-to-crisp membership settings. Borgelt shows that the regularization term which is derived from a maximum entropy approach yields desirable results which resemble the possibilistic clustering framework [15]. Nevertheless, utilization of a regularization term carries with it the need to properly configure the corresponding regularization coefficient, the value of which may be problem class or even problem instance dependent. The desired effect of that approach is that the data items are weighted in the update equation exactly according to their membership to the clusters. This is contrary to the conventional fuzzification regimes in which a function of the membership values is present in the update equations.

In (2.1), $m > 1$ is the *fuzzifier* (also called *weighing exponent* and *fuzziness*). The optimal choice for the value of the fuzzifier is a debated matter [16] and is suggested to be "an open question" [17]. Bezdek [18] suggests that $1 < m < 5$ is a proper range and utilizes $m = 2$. The use of $m = 2$ is suggested by Dunn [13] in his early work on the topic as well and also by Frigui et al. [19], among others [20]. Bezdek [21] provided physical evidence for the choice of $m = 2$ and Pal et al. [22] suggested that the best choice for m is probably in the interval $[1.5, 2.5]$. Yu et al. [17] argue that the choices for the value of m are mainly empirical and lack a theoretical basis. They worked on providing such a basis and suggested that "a proper m depends on the data set itself" [17]. Nevertheless, it is known that larger values of m soften the boundary between the clusters [14].

Recently, Zhou et al. [23] proposed a method for determining the optimal value of m in the context of FCM. They employed four Cluster Validity Index (CVI) models and utilized repeated clustering for $m \in [1.1, 5]$ on four synthetic data sets as well as four real data sets adopted from the UCI Machine Learning Repository [24] (refer to [25] for a review of CVIs and [26] for coverage in the context of relational clustering). The range for m in that work is based on previous research [27] which provided lower and upper bounds on m. The investigation carried in [23] yields that $m = 2.5$ and $m = 3$ are optimal in many cases and that $m = 2$ may in fact not be appropriate for an arbitrary set of data items. This result is in line with other works which demonstrate that larger values of m provide more robustness against noise and the outliers. Nevertheless, significantly large values of m are known to push the convergence towards the sample mean, in the context of Euclidean clustering [17]. Wu [28] analyzes FCM and some of its variants in the context of robustness and recommends $m = 4$. In [29], the authors address the related problem of fuzzy model construction. They set up the framework using $m = 2$ and proceed to find the optimal value of m for different problem instances while maintaining $m \in [1.1, 5]$. They show that different values of m are optimal within the context of different problem classes.

Rousseeuw et al. [30] suggested to replace f_{nc}^m with $\alpha f_{nc} + (1 - \alpha) f_{nc}^2$, for a known $0 < \alpha < 1$. Klawonn et al. [31, 32] suggested to generalize this effort and to replace f_{nc}^m with an increasing and differentiable function $g(f_{nc})$.

Pedrycz [33, 34, 35] suggested to modify (2.2) in favor of customized $\sum f_{nc}$ constraints for different values of n. That technique allows for the inclusion of *a priori* information into the clustering framework and is addressed as Conditional Fuzzy C-means (CFCM). The same modification is carried out in Credibilistic Fuzzy C-Means (CFCM) [36, 37], in which the "credibility" of data items is defined based on the distances between data items and clusters. Therefore, in that approach, (2.2) is modified in order to deflate the membership of outliers to the set of clusters (also see [38]). Customization of (2.2) is also carried out in Cluster Size Insensitive FCM (csiFCM) [39] in order to moderate the impact of data items in larger clusters on an smaller adjacent cluster. Leski [16] provides a generalized version of this approach in which $\sum \beta f_{nc}^\alpha$ is constrained.

2.2 Prototype-based Clustering

It is a common assumption that the notion of homogeneity depends on the distances between the data items. This assumption is made implicitly when clusters are modeled as *prototypical* data items, also called *clustroids* or cluster *centroids*, as in FCM, for example. A prominent choice in these works is the use of the Euclidean distance function [40]. For example, the potential function approach considers data items as energy sources scattered in a multi-dimensional space and seeks peak values in the field [41] (also see [42, 43, 44]). We argue, however, that the *distance* between the data items may not be either defined or meaningful and that what the clustering algorithm is to accomplish is the minimization of *data item-to-cluster* distances. For example, when data items are to be clustered into certain lower-dimensional subspaces, as it is the case with Fuzzy C-Varieties (FCV) [45], the Euclidean distance between the data items is irrelevant. We note that, in fact, fuzzy clustering is sometimes equated and reduced to prototype-based clustering [14].

Prototype-based clustering does not necessarily require prototypes which are explicitly present. For example, in kernel-based clustering, it is assumed that a non-Euclidean distance can be defined between any two data items. The clustering algorithm then functions based on an FCM-style objective function and produces clustroids which are defined in the same feature space as the data items [46]. These cluster prototypes may not be explicitly represented in the data item space, but, nevertheless, they share the same mathematical model as the data items [47] (the reader is referred to a review of Kernel FCM (KFCM) and Multiple-Kernel FCM (MKFCM) in [48] and several variants of KFCM in [49]). Another example for an intrinsically prototype-based clustering approach in which the prototypes are not explicitly "visible" is the Fuzzy PCA-guided Robust k-means (FPR k-means) clustering algorithm [50] in which a centroid-less formulation [51] is adopted which, nevertheless, defines homogeneity as proximity between the data items.

Relational clustering approaches constitute another class of algorithms which are intrinsically based on the distances between the data items (for example refer to Relational FCM (RFCM) [52] and its non-Euclidean extension Nerf C-means [53]). The goal of this class of algorithms is to group the data items into *self-similar* bunches. Another algorithm in which the presence of prototypes may be less evident is Multiple Prototype Fuzzy Clustering Model (FCMP) [54], in which data items are described as a linear combination of a set of prototypes, which are, nevertheless, members of the same $\mathbb{R}^k$ as the data items are. Additionally, some researchers utilize L_r-norms, for $r \neq 2$ [55, 56, 57, 58], or other distance functions which are defined between a pair of data items [59].

We argue that a successful departure from the assumption of prototypical clustering is achieved when clusters and data items have different mathematical models. For example, the Gustafson-Kessel algorithm [60] models a cluster as a pair of a point and a covariance matrix and utilizes the Mahalanobis distance between data items and clusters (also see the Gath-Geva algorithm [61]). Fuzzy shell clustering algorithms [20] utilize more generic geometrical structures. For example, the FCV [45] algorithm can detect lines, planes, and other hyper-planar forms, the Fuzzy C Ellipsoidal

Shells (FCES) [62] algorithm searches for ellipses, ellipsoids, and hyperellipsoids, and the Fuzzy C Quadric Shells (FCQS) [20] and its variants seek quadric and hyperquadric clusters (also see Fuzzy C Plano-Quadric Shells (FCPQS) [63]).

2.3 Robustification

Dave et al. [64] argue that the function of membership values in FCM and the concept of weight functions in robust statistics are related. Based on this perspective, it is argued that the classical FCM in fact provides an indirect means for attempting robustness. Nevertheless, it is known that FCM and other least square methods are highly sensitive to noise [36]. Hence, there has been ongoing research on the possible modifications of FCM in order to provide a (more) robust clustering algorithm [65, 66]. Dave et al. [64] provide an extensive list of relevant works and outline the intrinsic similarities within a unified view (also see [67, 68]).

The first attempt to robustifying FCM, based on one account [64], is the Ohashi Algorithm [67, 69]. That work adds a noise cluster to FCM and writes the robustified objective function as,

$$\Delta = \alpha \sum_{c=1}^{C} \sum_{n=1}^{N} f_{nc}^m \|x_n - \psi_c\|^2 + (1-\alpha) \sum_{n=1}^{N} \left(1 - \sum_{c=1}^{C} f_{nc}\right)^m. \tag{2.3}$$

The transformation from (2.1) to (2.3) was suggested independently by Dave [68, 70] when he developed the Noise Clustering (NC) algorithm as well. The core idea in NC is that there exists one additional imaginary prototype which is at a fixed distance from all of the data items and represents noise. That approach is similar to modeling approaches which perform consecutive identification and deletion of one cluster at a time [71, 72]. Those methods, however, are expensive to carry out and require reliable cluster validity measures.

Krishnapuram et al. [73] extended the idea behind NC and developed the Possibilistic C-means (PCM) algorithm by rewriting the objective function as,

$$\Delta = \sum_{c=1}^{C} \sum_{n=1}^{N} t_{nc}^m \|x_n - \psi_c\|^2 + \sum_{c=1}^{C} \eta_c \sum_{n=1}^{N} (1 - t_{nc})^m. \tag{2.4}$$

Here, t_{nc} denotes the degree of representativeness or *typicality* of x_n to ψ_c (also addressed as a *possibilistic degree* in contrast to the *probabilistic* model utilized in FCM). As expected from the modification in the way t_{nc} is defined, compared to that of f_{nc}, PCM removes the sum of one constraint, shown in (2.2), and in effect extends the idea of one noise cluster in NC into C noise clusters. In other words, PCM could be considered as the parallel execution of C independent NC algorithms that each seek a cluster. Therefore, the value of C is somewhat arbitrary in PCM [64]. For this reason, PCM has been called a *mode-seeking* algorithm where C is the upper bound on the number of modes.

We argue that the interlocking mechanism present in FCM, i.e. (2.2), is valuable in that, not only clusters seek homogenous sets, but that they are also forced into more optimal "positions" through forces applied by competing clusters. In other words, borrowing the language used in [40], in FCM, clusters "seize" data items and it is disadvantageous for multiple clusters to claim high membership to the same data item. There is no phenomenon, however, in NC and PCM which corresponds to this internal factor. Additionally, it is likely that PCM clusters coincide and/or leave out portions of the data unclustered [74]. In fact, it is argued that the fact that at least some of the clusters generated through PCM are non-coincidental is because PCM gets trapped into local minimum [75] (also see [14]). PCM is also known to be more sensitive to initialization than other algorithms in its class [40].

It has been argued that both concepts of possibilistic degrees and membership values have positive contributions to the purpose of clustering [76]. Hence, Pal et al. [77] combined FCM and PCM and rewrote the optimization function of Fuzzy Possiblistic C-Means (FPCM) as minimizing,

$$\Delta = \sum_{c=1}^{C} \sum_{n=1}^{N} \left(f_{nc}^m + t_{nc}^\eta \right) \|x_n - \psi_c\|^2, \qquad (2.5)$$

subject to (2.2) and $\sum_{n=1}^{N} t_{nc} = 1, \forall c$. That approach was later shown to suffer from different scales for f_{nc} and t_{nc} values, especially when $N \gg C$, and, therefore, additional linear coefficients and a PCM-style term were introduced to the objective function [78] (also see [79] for another variant). It has been argued that the resulting objective function employs four correlated parameters and that the optimal choice for them for a particular problem instance may not be trivial [40]. Additionally, in the new combined form, f_{nc} cannot necessarily be interpreted as a membership value [40].

Weight modeling is an alternative robustification technique and is exemplified in the algorithm developed by Keller [80], in which the objective function is rewritten as,

$$\Delta = \sum_{c=1}^{C} \sum_{n=1}^{N} f_{nc}^m u_c \frac{1}{\omega_n^q} \|x_n - \psi_c\|^2, \qquad (2.6)$$

subject to $\sum_{n=1}^{N} \omega_n = \omega$. Here, the values of ω_n are updated during the process as well.

Frigui et al. [19] included a robust loss function in the objective function of FCM and developed

Robust C-Prototypes (RCP),

$$\Delta = \sum_{c=1}^{C} \sum_{n=1}^{N} f_{nc}^m u_c \left(\|x_n - \psi_c\| \right). \tag{2.7}$$

Here, $u_c(\cdot)$ is the robust loss function for cluster c. They further extended RCP and developed an unsupervised version of RCP, nicknamed URCP [19]. Wu et al. [47] used $u_c(x) = 1 - e^{-\beta x^2}$ and developed Alternative HCM (AHCM) and Alternative FCM (AFCM) algorithms (also see [81]).

2.4 Number of Clusters and Cluster Fuzziness

The classical FCM and PCM, and many of their variants, are based on the assumption that the number of clusters is known (an extensive review of this topic is given in [10, Chapter 4]). While PCM-style formulations may appear to relax this requirement, the corresponding modification is carried out at the cost of yielding an ill-posed optimization problem [40]. Nevertheless, repeating the clustering for different numbers of clusters [61, 82] and Progressive Clustering are two of the alternative approaches to address the challenge of not requiring *a priori* knowledge about the number of clusters present in a particular set of data items.

Among the many variants of Progressive Clustering are methods which start with a significantly large number of clusters and freeze "good" clusters [82, 83, 63], approaches which combine compatible clusters [84, 82, 83, 63, 19], and the technique of searching for one "good" cluster at a time until no more is found [71]. Use of regularization terms in order to push the clustering results towards the "appropriate" number of clusters is another approach taken in the literature [85]. These regularization terms, however, generally involve additional parameters which are to be set *carefully*, and potentially per problem instance [77]. Moreover, while for many cluster models merging two clusters may not be applicable, defining a cluster validity model which is independent of the particular cluster model of an arbitrary problem class may not be trivial.

Dave and Krishnapuram conclude in their 1997 paper that the solution to the general problem of robust clustering when the number of clusters is unknown is "elusive" and that the techniques available in the literature each have their limitations [64].

2.5 Weighted Clustering

Many fuzzy and possibilistic clustering algorithms make the assumption that the data items are equally important. Weighted fuzzy clustering, however, works on input data items which have an associated positive weight [71]. This notion can be considered as a marginal case of clustering fuzzy data [86]. Other examples for this setting include clustering of a weighted set, clustering of sampled data, clustering in the presence of multiple classes of data items with different priorities [87], and a measure used in order to speed up the execution through data reduction [88, 89, 90, 91]. Nock et al. [92] formalize the case in which weights are manipulated in order to move the clustering results towards data items which are harder to include regularly. Chen et al. [93] utilize density motivated weights in order to reduce the impact of outliers (refer to [94] for different variants of this framework). Semi Supervised FCM (ssFCM) [95] uses weight factors based on an Euclidean norm in order to balance the sizes of different hyper-spherical shaped clusters based on user intervention. Note that the extension of FCM on weighted sets has been developed under different names, including Density-Weighted FCM (WFCM) [90], Fuzzy Weighted C-means (FWCM) [96], and New Weighted FCM (NW-FCM) [97].

Chapter 3

Developed Method

3.1 Design Methodology

In this work, we utilize the two concepts of *Problem Class* and *Problem Instance* in order to develop a class-independent clustering algorithm. In this context, a problem class, as the name implies, defines a particular class of clustering problems. In other words, a problem class is an abstract mathematical object which defines specific mathematical models for data items as well as for clusters and their relationship. Hence, implied in the definition of a problem class are a distance function and a homogeneity model. A problem instance, on the other hand, is one realization of a particular problem class. For example, segmentation of color data into homogenous sets is a problem class and the segmentation of a particular image is a problem instance.

A main contribution of this paper is to provide a solution strategy to a generic definition of data clustering which is applicable to many problem classes. The method described in this paper is designed so that the utilization of the proposed method for a particular problem class follows a number of deterministic steps and that the execution of the proposed method for a particular problem instance does not require user intervention.

Figure 3.1 provides a pictorial representation of the interactions between the method developed in this paper and the two concepts of problem class and problem instance. Utilizing the language of software engineering, the method developed in this paper, which we have nicknamed *Selma*, accepts three function pointers and a scale parameter. Hence, in order for a particular problem class to adopt the developed method, it is required to provide appropriate instances of these functions and to set the scale parameter. Note that the three functions as well as the scale parameter are unique for different problem instances within a problem class. This process is equivalent to class

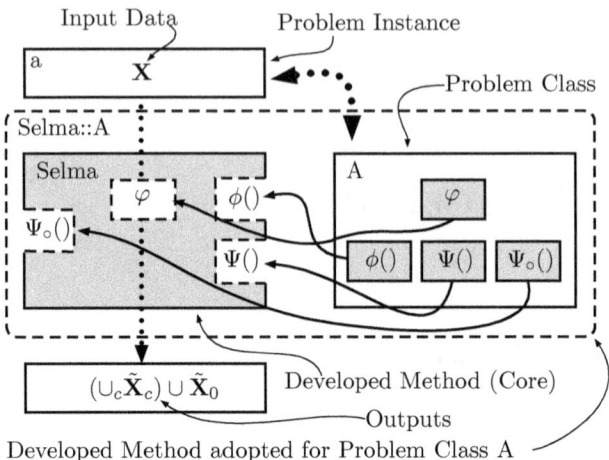

Figure 3.1: Pictorial representation of the concepts of *Problem Class* and *Problem Instance* in the context of the present work and their relationship with the method developed in this paper.

inheritance in software engineering, in which, from the design perspective, different adoptions of the method proposed in this work directly inherit the core processes, which are class independent, from the parent class *Selma* and override certain class methods. Additionally, problem classes are responsible for input preparation and output presentation.

3.2 Model Preliminaries

We assume that any problem class provides a mathematical model for data items and denote a data item as x. For example, a particular problem class may utilize data items which are modeled as $x \in \mathbb{R}^k$, for a known k. We also assume that any problem class defines a cluster model, which complies with the *notion of homogeneity* relevant to the problem class at hand, and denote a cluster as ψ. Hence, for example, for the aforementioned data item model $x \in \mathbb{R}^k$, and assuming that *Euclidean compactness* is the target, the problem class may also define $\psi \in \mathbb{R}^k$. In this context, Euclidean compactness is the notion of homogeneity applicable to this arbitrary problem class. Note that a problem class generally relates to a physical phenomenon which mandates a specific relationship between the data items and the clusters. In this work, we model this relationship as the three functions $\phi()$, $\Psi()$, and $\Psi_\circ()$ and the scale parameter φ. These identities are schematically presented in Figure 3.1 and are introduced formally later in this section. This modeling framework has precedence in the literature [98, 99].

In this work, we utilize a weighted set of data items, defined as,

$$\mathbf{X} = \left\{ (\omega_n; x_n) \right\}, n = 1, \cdots, N, \omega_n > 0, \tag{3.1}$$

and we define the *weight* of $\mathbf{X}$ as $\Omega(\mathbf{X}) = \sum_{n=1}^{N} \omega_n$. When known in the context, we abbreviate $\Omega(\mathbf{X})$ as Ω. Thus, when estimating expected values, we treat $\mathbf{X}$ as a set of realizations of the random variable x and write,

$$p\{x_n\} = \frac{\omega_n}{\Omega}. \tag{3.2}$$

Hence, this framework models fuzziness of the input data through weights associated with the data items.

We assume that the real-valued positive *distance function* $\phi(x, \psi)$ is defined. Through this abstraction, we allow for any applicable distance functions and therefore decidedly avoid the dependence of the underlying algorithm on Euclidean or any other particular notations of distance. We also assume that the distance function is unbounded, i.e. that for any cluster representation ψ and any positive value L, there exist infinite number of data items x for which $\phi(x, \psi) > L$. As examples, when the data items belong to $\mathbb{R}^k$, the Euclidean Distance, any ℓ_p norm, and the Mahalanobis Distance are special cases of the notion of data item to cluster distance defined here. The corresponding cluster models in these cases denote $\psi \in \mathbb{R}^k$, $\psi \in \mathbb{R}^k$, and ψ identifying a pair of a member of $\mathbb{R}^k$ and a $k \times k$ covariance matrix, respectively.

We assume that $\phi(x, \psi)$ is differentiable in terms of ψ and that for any non-empty weighted set $\mathbf{X}$, the following function of ψ,

$$\Delta_{\mathbf{X}}(\psi) = E\{\phi(x, \psi)\} = \frac{1}{\Omega} \sum_{n=1}^{N} \omega_n \phi(x_n, \psi), \tag{3.3}$$

has one and only one minimizer which is also the only solution to the following equation,

$$\sum_{n=1}^{N} \omega_n \frac{\partial}{\partial \psi} \phi(x_n, \psi) = 0. \tag{3.4}$$

In this paper, we assume that a function $\Psi(\cdot)$ is given, which, for the input weighted set $\mathbf{X}$, produces the optimal ψ which minimizes (3.3) and is the solution to (3.4). We address $\Psi(\cdot)$ as the *cluster fitting function*. Examples for $\Psi(\cdot)$ include the mean and the median when x and ψ are real values and $\phi(x, \psi) = (x - \psi)^2$ and $\phi(x, \psi) = |x - \psi|$, respectively.

Note that $\Psi(\cdot)$ is a solution to the M-estimator given in (3.3). We note that when a closed-form representation for $\Psi(\cdot)$ is not available, conversion to a W-estimator can produce a procedural solution to (3.4) [100]. Additionally, many of the techniques developed in the context of Weber Problems [101, 102] may be applicable to finding a procedural solution to $\Psi(\cdot)$. The reader is referred to Weiszfeld's algorithm [103, 104], Iteratively Reweighted Least Squares (IRLS) [105], and the methods developed by Tikhonov-Arsenin [106], Levenberg [107], and Marquardt [108] (also see [109]). A very recent review of the different incarnations of the Weber Problem and an outline of Weiszfeld's work can be found in [110].

We assume that a function $\Psi_\circ(\cdot)$, which may depend on $\mathbf{X}$, is given, that produces a requested number of initial clusters. We address this entity as the *cluster initialization function* and denote the number of clusters produced by it as C. In this work we utilize cluster fuzziness in order to allow the algorithm to converge to a number of clusters which is appropriate for the input set of data items. Hence, the number of clusters generated by $\Psi_\circ(\cdot)$ is in effect the starting point for the developed algorithm.

We assume that a robust loss function, $u(\cdot) : [0, \infty] \to [0, 1]$, is given which satisfies $\lim_{\tau\to\infty} u(\tau) = 1$. Additionally, we assume that $u(\cdot)$ is an increasing differentiable function which satisfies $u(0) \simeq 0$ and $u(1) \simeq \frac{1}{2}$.

In this work, we utilize the rational robust loss function given below,

$$u(x) = \frac{x + \varepsilon}{1 + x}, 0 < \varepsilon \ll 1. \tag{3.5}$$

Here, in order to avoid numerical instability at $x = 0$, $\varepsilon = 2 \times 10^{-16}$ is utilized. Derivation shows that $u'(x) = (1 - \varepsilon)(1 + x)^{-2}$ and $u''(x) = -2(1 - \varepsilon)(1 + x)^{-3}$, i.e. $u(x)$ is an increasing concave function.

Using (3.5), we model the loss of x_n when it belongs to ψ_c as,

$$u_{nc} = u\left(\frac{1}{\lambda}\phi_{nc}\right), \phi_{nc} = \phi(x_n, \psi_c). \tag{3.6}$$

Additionally, we model the loss of a data item which is considered to be an outlier as the positive constant U. In Section 3.8, we propose two procedures which determine the values of λ and U for any given problem class.

We address λ as the *scale* parameter (note the similarity with the cluster-specific weights in PCM [73]). In fact, λ has a similar role to that of scale in robust statistics [111] (also called the *resolution* parameter [42]) and the idea of distance to noise prototype in the NC algorithm [68, 70].

Scale can also be considered as the controller of the boundary between inliers and outliers [64]. From a geometrical perspective, λ controls the radius of spherical clusters and the thickness of planar and shell clusters [40]. One may investigate the possibility of generalizing this unique scale factor into cluster-specific scale factors, i.e. λ_c values, in line with the η_c variables in PCM [73].

We acknowledge that one may consider the possibility of utilizing Tukey's biweight [112], Hampel [113], and Andrews loss functions within the framework developed in this paper. Moreover, the Blake-Zisserman [114] loss function may be a proper choice for the present work because in addition to being bounded, it is based on a Gaussian model for the inliers and a uniform one for the outliers. Huber and Cauchy loss functions, however, are not bounded and therefore are not applicable to this work. Refer to [64, Table I] for mathematical formulations.

3.3 Cluster Fuzziness

As briefly reviewed in Chapter 1sec:theory:count, one of the contributions of this work is the introduction of the concept of cluster fuzziness. In order to define this concept, we assume that a particular clustering solution comprises of the C clusters $\psi_1, \cdots, \psi_C$. Note that in the literature, these clusters are in fact crisp entities. In other words, not only after the convergence, but at any time during the clustering process, the C clusters $\psi_1, \cdots, \psi_C$ are non-fuzzy entities. In this work, we challenge this assumption and work with clusters which are defined in a fuzzy context.

We propose to model the probability that ψ_c is relevant, as $0 \leq r_c \leq 1$, and address it as the *relevance* of ψ_c. This model changes the dichotomy of "ψ_c exists/does not exist" into "ψ_c is relevant with the probability of r_c", i.e. $p\{\exists \psi_c\} = r_c$. We note that the transformation of the clusters from crisp entities, which exist within a binary context, into fuzzy ones, with associated relevance probabilities, is a reincarnation of the transformation which fuzzifies the relationship between the data items and the clusters. In other words, this work adopts the transformation which converted hard clustering to fuzzy clustering and extends it to not only include the data items but also to govern the relevance of the clusters. Moreover, cluster fuzziness, as developed in this work, transforms the concept of *number of clusters* from a crisp notion into a fuzzy one, as will be discussed next.

We assume that at a certain state during a fuzzy clustering process, it is deemed that the C clusters $\psi_1, \cdots, \psi_C$ are present in the input set of data items. This statement, in the conventional fuzzy clustering literature, translates into the number of clusters being equal to C. In the context of the present work, however, the number of clusters is in fact a non-deterministic value which is not necessarily an integer either. To demonstrate this distinction, we assume that the relevance factor for the aforementioned clusters are denoted as $r_1 \cdots r_C$. In this context, the *expected* number of clusters in the system at the aforementioned state is calculated as follows,

$$E\{\mathbf{C}\} = \sum_{c=1}^{C} p\{\exists \psi_c\} = \sum_{c=1}^{C} r_c. \tag{3.7}$$

The present work not only modifies the mathematical model for the concept of C, but also it defines three facets for it. In conventional fuzzy clustering, the number of clusters in the system is denoted as C. This value may in fact change during the execution, due to a pruning process for example, but, nevertheless, the "Number of Clusters" entity is a singleton. In the proposed method, however, we utilize the three entities of C, $E\{\mathbf{C}\}$ and $C_\circ$, to address different aspects of a comparable concept, as described next.

We denote the number of clusters *present* in the system at any time during the execution of the algorithm as C. Note that the r_c value corresponding to some of these clusters may in fact be zero. Thus, the second descriptor for the number of clusters present in the system, i.e. $E\{\mathbf{C}\}$, is always less than or equal to C. Here, $E\{\mathbf{C}\}$ denotes the *expected* number of clusters present in the system, as derived in (3.7). In the present work, C and $E\{\mathbf{C}\}$ are expected to approach $C_\circ$, the *ideal* number of clusters. This is the value that is given by the user, based on user experience or other *a priori* sources of information.

In this work, the clustering algorithm starts with $C = \kappa C_\circ$ and goes through a process in which C is always non-increasing. We address κ as the *cluster redundancy factor* and maintain $\kappa \geq 1$. We note that the important factor in selecting the value of κ is the processing power budget available to the algorithm. We explore this concept in Chapter 4 in more detail.

While κ needs to be equal or greater than 1, we argue that $\kappa \geq 2$ is a more optimal choice. In fact, $2C_\circ$ is the maximum number of clusters which satisfy $r_c \geq \frac{1}{2}, \forall c$ and $E\{\mathbf{C}\} = C_\circ$. In other words, assuming that every cluster is "on the brink of relevance", a set $\mathbf{X}$ which is expected to contains $C_\circ$ clusters can at most accommodate $2C_\circ$ clusters. Nevertheless, one many choose larger values of κ with the knowledge that some of the clusters will in fact be removed during the clustering process. The logic for selecting a large value of κ is to allow the search space to grow, and thus to expect that spurious clusters will give way to clusters which are relevant to $\mathbf{X}$.

It must be noted that we in fact *loosely enforce* $E\{\mathbf{C}\} = C_\circ$ during the process developed in this paper. As will be discussed later in this section, the proposed method in fact enforces $E\{\mathbf{C}\} = C_\circ$

but it also requires $0 \leq r_c \leq 1$. This secondary requirement can in fact modify the value of the r_cs in a way that $E\{\mathbf{C}\} = C_\circ$ may not be satisfied anymore. We address this aspect of the proposed method in more details later in this section as well as in Chapter 4.

Hence, we start with $C = \kappa C_\circ$, as produced by $\Psi_\circ(\cdot)$, and set $r_c = 1, \forall c$. Then, we allow the r_c values to change as the system moves towards minimizing the loss. Through this process, some of the clusters will fade away, i.e. some of the r_c values will be less than $\frac{1}{2}$, and thus the corresponding clusters will be removed from the pool of clusters. Note that, the number of clusters which survive this process is non-deterministic, with the stipulation that $C \simeq C_\circ$ is the estimated number of clusters. The details of this process are outlined later in this paper.

3.4 Assessment of Loss

We assume that the developed clustering procedure has found C clusters in the input set of data items $\mathbf{X}$ and address them as $\psi_1, \cdots, \psi_C$. We also assume that a Maximum Likelihood procedure has been applied on $\mathbf{X}$ and denote the set of data items which are assigned to ψ_c as $\tilde{\mathbf{X}}_c$. We address the union of all $\tilde{\mathbf{X}}_c$ for $c = 1, \cdots, C$ as $\tilde{\mathbf{X}}$. In this context, the set $\tilde{\mathbf{X}}_0 = \mathbf{X} - \tilde{\mathbf{X}}$ contains the data items which are considered to be outliers. Note that, here, when using set operators such as union and subtraction for weighted sets, we drop the weights. In the following, when we use the term "clustering solution", we refer to a particular set of clusters and a particular relationship between the data items and the clusters.

In this paper, we assume that an arbitrary clustering solution is given and assign to it a non-negative loss value through Bayesian inference. The purpose of the clustering algorithm developed in this paper is in fact the minimization of this loss, as will be described later. In order to arrive at the model for loss, we consider an arbitrary data item x_n. This data item may be an outlier or it may belong to one of the C clusters. Hence, we model the loss associated with x_n, for a particular clustering solution, as follow.

$$E\{Loss|x_n\} = p\{x_n \in \tilde{\mathbf{X}}_0\} E\{Loss|x_n \in \tilde{\mathbf{X}}_0\} + \quad (3.8)$$
$$p\{x_n \in \tilde{\mathbf{X}}\} \sum_{c=1}^{C} p\{x_n \in \tilde{\mathbf{X}}_c | x_n \in \tilde{\mathbf{X}}\} E\{Loss|x_n \in \tilde{\mathbf{X}}_c\}.$$

In assessing the loss for a data item given that it belongs to a cluster, we consider the probability that that certain cluster is in fact relevant, as follows,

$$E\{Loss|x_n\} = p\left\{x_n \in \tilde{\mathbf{X}}_0\right\} E\left\{Loss|x_n \in \tilde{\mathbf{X}}_0\right\} \quad (3.9)$$

$$+ p\left\{x_n \in \tilde{\mathbf{X}}\right\} \sum_{c=1}^{C} p\left\{x_n \in \tilde{\mathbf{X}}_c | x_n \in \tilde{\mathbf{X}}\right\} \left[p\{\exists \psi_c\} E\left\{Loss|x_n \in \tilde{\mathbf{X}}_c \,\&\, \exists \psi_c\right\} \right.$$

$$\left. + p\{\nexists \psi_c\} E\left\{Loss|x_n \in \tilde{\mathbf{X}}_c \,\&\, \nexists \psi_c\right\} \right].$$

We now denote the probability that x_n is an inlier as p_n and the probability that x_n belongs to $\tilde{\mathbf{X}}_c$ given that it is an inlier as f_{nc} and rewrite (3.9) as follows.

$$E\{Loss|x_n\} = (1-p_n)E\left\{Loss|x_n \in \tilde{\mathbf{X}}_0\right\} + p_n \sum_{c=1}^{C} f_{nc} \Big[\quad (3.10)$$

$$r_c E\left\{Loss|x_n \in \tilde{\mathbf{X}}_c \,\&\, \exists \psi_c\right\} + (1-r_c) E\left\{Loss|x_n \in \tilde{\mathbf{X}}_c \,\&\, \nexists \psi_c\right\} \Big].$$

As seen in (3.10), in order to assess loss in the system, we need to model the loss associated with an inlier which is assigned to a relevant cluster, i.e. $E\left\{Loss|x_n \in \tilde{\mathbf{X}}_c \,\&\, \exists \psi_c\right\}$, as well as two other cases. These latter cases correspond to a data item which does not belong to any of the clusters, i.e. $E\left\{Loss|x_n \in \tilde{\mathbf{X}}_0\right\}$, and a data item which belongs to non-relevant cluster, i.e. $E\left\{Loss|x_n \in \tilde{\mathbf{X}}_c \,\&\, \nexists \psi_c\right\}$. Here, we model the two latter cases as the constant "outlier loss" U. In Section 3.8 we propose a deterministic process for setting the value of U for any particular problem class.

Now, we utilize (3.6) and rewrite (3.10) as follows,

$$E\{Loss|x_n\} = (1-p_n)U + p_n \sum_{c=1}^{C} f_{nc}\left(r_c u_{nc} + (1-r_c)U\right). \quad (3.11)$$

Now, we utilize (3.11) in order to assess the loss corresponding to the entire set $\mathbf{X}$, as follows,

$$E\{Loss|\mathbf{X}\} = \sum_{n=1}^{N} p\{x_n\} E\{Loss|x_n\} = \frac{1}{\Omega} \sum_{n=1}^{N} \omega_n \left[p_n \sum_{c=1}^{C} f_{nc} \quad (3.12) \right.$$

$$\left. (r_c u_{nc} + (1-r_c)U) + (1-p_n)U \right].$$

Subsequently, we drop the constant Ω from (3.12) and write the objective function for the clustering problem developed in this paper as follows.

$$\Delta = \sum_{n=1}^{N}\sum_{c=1}^{C} \omega_n p_n f_{nc} r_c u_{nc} + \sum_{n=1}^{N}\sum_{c=1}^{C} \omega_n p_n f_{nc}(1-r_c)U + \sum_{n=1}^{N} \omega_n(1-p_n)U 2C \frac{1}{2C}. \quad (3.13)$$

Close assessment of (3.13) shows that this cost function complies with an HCM-style hard template. It is known, however, that the utilization of the concept of the fuzzifier has important benefits, as outlined in Section 2.1. Hence, we rewrite (3.13) and derive the following objective function for the clustering problem proposed in this paper.

$$\Delta = \sum_{n=1}^{N}\sum_{c=1}^{C} \omega_n p_n^m f_{nc}^m r_c^m u_{nc} + \sum_{n=1}^{N}\sum_{c=1}^{C} \omega_n p_n^m f_{nc}^m (1-r_c)^m U + \qquad (3.14)$$
$$\sum_{n=1}^{N} \omega_n (1-p_n)^m U (2C)^{1-m}.$$

This objective function is to be minimized subject to (2.2) and,

$$\sum_{c=1}^{C} r_c = C_\circ. \qquad (3.15)$$

In Section 3.5 we provide a solution strategy for this optimization problem.

In (3.13) we have chosen to write U as $U2C\frac{1}{2C}$ in order to compensate for the impact caused by the incorporation of the fuzzifier into the cost function. In fact, three types of terms exist in (3.13), i.e. $p_n f_{nc} r_c$, $p_n f_{nc}(1-r_c)$, and $(1-p_n)$. These terms are each products of membership identifiers. In other words, for example, the term $p_n f_{nc} r_c$ denotes the probability that x_n is an inlier which belongs to ψ_c which is a relevant cluster. However, while the first two terms contain three elements, the third one only contains the single element p_n. We argue that this is because this term in fact contains implicit components of the type *"if either P or not P"* hidden in it. In other words, the cost component $U(1-p_n)$ in fact models the situation in which x_n is an outlier, in which case it is irrelevant whether or not x_n belongs to any of the clusters and whether or not the said clusters are relevant or not. In other words, the term $(1-p_n)$ is in fact the simplified version of the following term,

$$(1-p_n)\Big[\sum_{c=1}^{C} f_{nc}\Big]\Big[r_c + (1-r_c)\Big] = 1 - p_n. \qquad (3.16)$$

While this alternative form is in effect identical to $1 - p_n$, the difference becomes significant when the fuzzifier is integrated into the objective function. In fact, with the addition of the fuzzifier, the term given in (3.16) ought to be modified to,

$$(1-p_n)^m \Big[\sum_{c=1}^{C} f_{nc}^m\Big]\Big[r_c^m + (1-r_c)^m\Big] \leq (1-p_n)^m. \qquad (3.17)$$

In other words, if no other measure is taken, the incorporation of the fuzzifier will effectively reduce the cost of being an outlier, as explained below.

We note that for any set of K non-negative variables ζ_k which satisfy $\sum_{k=1}^{K} \zeta_k = 1$, we have $\sum_{k=1}^{K} \zeta_k^m \leq 1$, when $m > 1$. The equality in this relationship, i.e. the upper bound, occurs when all of the ζ_k are zero except for one which is unity. The lower bound on $\sum_{k=1}^{K} \zeta_k^m$, however, occurs when the ζ_k are identical. Therefore, we replace (3.16) with the case in which all the f_{nc} are equal and $r_c = 1 - r_c$. This process guarantees that when the fuzzifier is incorporated into the cost function, the corresponding term is always greater than or equal to the pre-fuzzifier term. In other words, we replace $(1-p_n)$ with $(1-p_n)C\frac{1}{C}2\frac{1}{2}$ and therefore, after the incorporation of the fuzzifier, yield $(1-p_n)^m C\frac{1}{C^m}2\frac{1}{2^m} = (1-p_n)^m(2C)^{1-m}$. Above, this transformation was rephrased, *imprecisely*, as substituting U with $U2C\frac{1}{2C}$ in (3.13). We revisit the implications of this substitution in Section 3.8 when we discuss the process for determining U and λ.

Before addressing the solution strategy for (3.14), we note that, as discussed in Section 2.1, there is consensus in a part of the community that $m = 2$ is one of the appropriate choices for the fuzzifier. Hence, we rewrite (3.14) as,

$$\Delta = \sum_{n=1}^{N}\sum_{c=1}^{C} \omega_n p_n^2 f_{nc}^2 r_c^2 u_{nc} + \sum_{n=1}^{N}\sum_{c=1}^{C} \omega_n p_n^2 f_{nc}^2 (1-r_c)^2 U + \sum_{n=1}^{N} \omega_n (1-p_n)^2 \frac{U}{2C}. \qquad (3.18)$$

3.5 Solution Strategy

The objective function derived in Section 3.4 is to be minimized through finding the optimal values for p_n, f_{nc}, and r_c values as well as ψ_c. In this section, we develop an Alternating Optimization framework which modifies some of these variables in each step towards a potentially local optimal solution. Note that the formulations developed in this section are in fact too expensive to be utilized in the actual implementation. Hence, in Section 3.9 we develop auxiliary variables which increase the efficiency of this process from the vantage point of resource requirement.

First, we optimize (3.18) subject to f_{nc}. To do so, we utilize Lagrange Multipliers and incorporate (2.2) into (3.18) and derive,

$$f_{nc} = \frac{\left(r_c^2 u_{nc} + (1-r_c)^2 U\right)^{-1}}{\sum_{c'=1}^{C} \left(r_{c'}^2 u_{nc'} + (1-r_{c'})^2 U\right)^{-1}}. \qquad (3.19)$$

We then calculate the partial derivative of Δ relative to p_n and use (3.19) and derive,

$$p_n = \frac{1}{1 + \dfrac{C}{\sum_{c=1}^{C}\left(r_c^2 u_{nc} + (1-r_c)^2 U\right)^{-1}}}. \tag{3.20}$$

Note that both (3.19) and (3.20) only depend on r_c, and not on f_{nc} or p_n.

Then, we utilize Lagrange Multipliers and incorporate (3.15) into (3.18) and calculate the partial derivatives of Δ relative to r_c and subsequently equate them to zero in order to derive,

$$r_c^\star = \frac{1}{\sum_{n=1}^{N} \omega_n p_n^2 f_{nc}^2 (u_{nc} + U)} \cdot \frac{C_\circ - \sum_{c'=1}^{C} \dfrac{\sum_{n=1}^{N} \omega_n p_n^2 f_{nc'}^2 U}{\sum_{n=1}^{N} \omega_n p_n^2 f_{nc'}^2 (u_{nc'} + U)}}{\sum_{c'=1}^{C} \dfrac{1}{\sum_{n=1}^{N} \omega_n p_n^2 f_{nc'}^2 (u_{nc'} + U)}} + \frac{\sum_{n=1}^{N} \omega_n p_n^2 f_{nc}^2 U}{\sum_{n=1}^{N} \omega_n p_n^2 f_{nc}^2 (u_{nc} + U)}. \tag{3.21}$$

Note that there is no guarantee that the $r_c^\star$ calculated in (3.21) will belong to $[0,1]$. In fact, during experiments we have witnessed instances in which either $r_c^\star < 0$ or $r_c^\star > 1$. Hence, we utilize,

$$r_c = \begin{cases} 0 & r_c^\star < 0 \\ 1 & r_c^\star > 1 \\ r_c^\star & Otherwise \end{cases}. \tag{3.22}$$

Nevertheless, this process can void (3.15). We opt to accept this sidestep. This deviation is in fact the reason we stated in Section 3.3 that (3.15) is *loosely enforced*.

Finally, we equate the partial derivative of Δ relative to ψ_c to zero and derive,

$$\psi_c = \Psi\left(\left\{(\omega_{nc}; x_n)\right\}\right). \tag{3.23}$$

Here,

$$\omega_{nc} = \frac{\omega_n p_n^2 f_{nc}^2}{(\phi_{nc} + \lambda)^2}. \tag{3.24}$$

3.6 Outlier Detection and Classification

The relationship between x_n and ψ_c is affected by both f_{nc} and p_n. We emphasize, however, that this situation is characteristically different from FPCM, where a clear notion of membership to cluster cannot be inferred from the variables involved in the model [40]. In the present work, however, model variables directly relate to the level of membership of each data item to each cluster and whether or not each data item is more likely to be an outlier. We develop these relationships in this section. Additionally, we develop a process which classifies the data items into disjoint hard sets.

Taking the arbitrary data item x_n, we recognize that p_n denotes the probability that x_n is an inlier. Hence, if $p_n < \frac{1}{2}$, we add x_n to the set of outliers, which we denote as $\tilde{\mathbf{X}}_0$. Otherwise, we include x_n in $\tilde{\mathbf{X}}_{c_n}$. Here, the integer $1 \leq c_n \leq C$ denotes the cluster for which f_{nc} is the largest (i.e. $c_n = \arg_c \max\{f_{nc}\}$). For the purpose of coherence, we define $c_n = 0$ for outliers.

3.7 Cluster Maintenance

As presented in Section 3.4, in this paper, we associate the fuzzy relevance indicator r_c to ψ_c. We use these values in order to decide if a cluster must remain in the solution or not. To do so, We allow the optimization process to converge and then eliminate the cluster which corresponds to the smallest value of r_c which is less than $\frac{1}{2}$. We then repeat the clustering iterations and continue the execution until the optimization process converges and no cluster can be eliminated.

3.8 Determination of U and λ

It is evident that λ defines the scale for ϕ_{nc}. This is exemplified in (3.6) and also everywhere else in this paper where ϕ_{nc} is divided by λ. We argue that, similarly, U defines the scale for u_{nc}. Hence, we argue that the two identities ϕ_{nc} and u_{nc} are *brought into context* through λ and U, respectively. We use this perceptual definition in order to propose a procedure for determining the appropriate values for λ and U for a particular problem class.

We suggest an imaginary situation, as depicted in Figure 3.2, in which two data items interact with two clusters ($N = 2$ and $C = 2$). The first data item, here x_1, is infinitely far from both clusters, in which case we expect $f_{11} = f_{12} = \frac{1}{2}$. This equilibrium is in fact verified by (3.19). The second data item, here x_2, however, is infinitely far from the second cluster and is at the distance

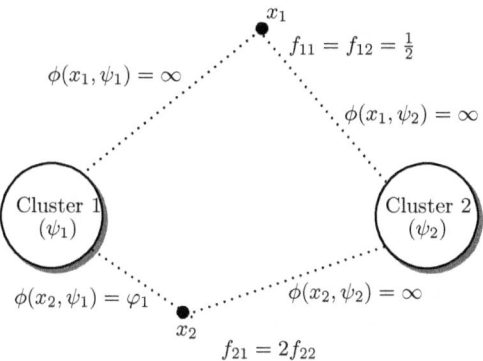

Figure 3.2: The process of selecting φ_1, based on which the value of λ is determined.

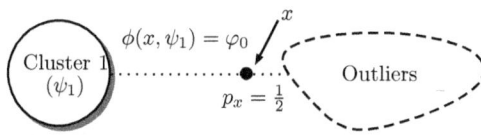

Figure 3.3: The process of selecting φ_0, based on which the value of U is determined.

of φ_1 from the first one. Here, we ask what value of φ_1 will result in $f_{21} = 2f_{22}$. This question can be asked in a different setting as follows: For a data item which is infinitely far from two clusters, how close should it get to one cluster, while maintaining its distance to the other, in order to be favored by the former cluster two times than the latter?

Utilizing (3.19) one can show that,

$$\lambda = \frac{1}{u^{-1}\left(2^{-(m-1)}\right)} \varphi_1. \tag{3.25}$$

Considering the special case utilized in the experiments carried in this paper, i.e. $m = 2$, one can show that independent of the choice of $u(\cdot)$, (3.25) leads to $\lambda \simeq \varphi_1$.

In order to estimate the proper value for U, we utilize another imaginary situation in which one data item interacts with one cluster, as depicted in Figure 3.3. Here, we ask how far the data item should be from the cluster in order for it to be an outlier with a probability of half? This situation, in effect, defines the boundary of inliers and outliers when Maximum Likelihood is applied on p_n. We denote this distance as φ_0 and derivation using (3.20) shows that,

$$U = 2u\left(\frac{1}{\lambda}\varphi_0\right). \tag{3.26}$$

It is important to emphasize that the outcome of this process is independent of C. In other words, in a set of data items which contains C clusters, a data item which is at the same distance of φ_0 from all of them will be an inlier with the probability of half. This is the result of the decision to replace U with $U2C\frac{1}{2C}$ in (3.13). In fact, in the absence of this substitution, (3.26) changes to the following equation,

$$U = \frac{1}{C}u\left(\frac{1}{\lambda}\varphi_0\right). \tag{3.27}$$

In this equation, U depends on C. In other words, the cost associated to an outlier depends on the number of clusters present in the system. We find this dependence hard to understand from a perceptual standpoint. The utilized formulation, i.e. the one given in (3.26), however, yields the case in which U is a property of the problem class and addition or removal of a cluster does not affect it.

We note that φ_0 and φ_1 reflect different aspects of the relationship between a data item and a cluster. Nevertheless, conceptually, we can envision that one may want to set $\varphi_0 = \varphi_1 = \varphi^2$. In this line of reasoning, φ^2 denotes the territory of a cluster, within which the cluster considers a data item as an inlier, therefore $p_x \geq \frac{1}{2}$, and also owns the data item when in competition with another farther cluster, therefore $f_{cx} \geq \frac{1}{2}$. Note that we use φ^2 instead of φ due to the fact that many distance functions are in fact the square of a geometrical entity. We arrived at the same point empirically as well, i.e. as will be discussed later, setting $\varphi_0 = \varphi_1$ *makes sense* in practice.

Reworking (3.25) for $\varphi_0 = \varphi_1 = \varphi^2$ we arrive at $\lambda \simeq \varphi^2$, as expected. The case of U, however, is more interesting. Substituting $\varphi_0 = \varphi_1 = \varphi^2$ in (3.26) we derive $U = 1$, independent of φ. This result corresponds to the case of $m = 2$.

We note that one may consider an alternative process in which the values for λ and/or U are determined based on one or more given sets of data items which are known to be homogenous.

3.9 Implementation Notes

Figure 3.4 shows the relationship between the variables which constitute a clustering solution. As it is the case for any Alternating Optimization flow, these variables alter during the process until convergence is achieved. In fact, the arrows in Figure 3.4 exhibit which variable impacts which one during this procedure. Note that in a grand scheme, every variable impacts every other one, except for Δ which is a function of every other variable. In fact, the process is deemed to have converged when relative change in Δ is negligible. This is visualized in Figure 3.4 through the two loops

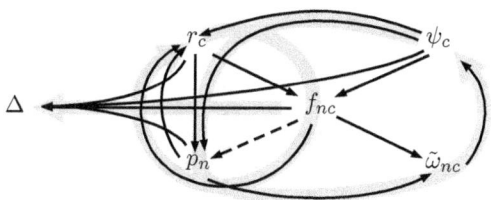

Figure 3.4: Flow of the variables in the Alternating Optimization process developed in this paper.

and the collective outpouring into Δ. In short, the two "wheels" of the algorithm revolve until the measure of convergence, Δ, settles. We note that the convergence test should in fact examine whether the clusters have become static. However, in the general case, the examination of cluster variability is non-trivial, and, therefore, we capitalize on the variability of the cost function as a sign that the solution to the clustering problem is still evolving.

While the flow shown in Figure 3.4 and the update equations given in Section 3.5 provide the basic structure of the developed method, they are not optimal for implementation. Hence, in this section, we provide auxiliary variables which facilitate the iterative application of the calculations developed in Section 3.5.

First, we define,

$$v_{nc} = \frac{1}{r_c^2 u_{nc} + (1-r_c)^2 U}, v_n = \sum_{c=1}^{C} v_{nc}. \tag{3.28}$$

Now, we rewrite (3.19) and (3.20) as,

$$f_{nc} = \frac{1}{v_n} v_{nc}, p_n = \frac{1}{1 + 2CU^{-1}\frac{1}{v_n}}. \tag{3.29}$$

Then, we define,

$$q_{nc} = \omega_n p_n^2 f_{nc}^2, q_c = U \sum_{n=1}^{N} q_{nc}, \omega_c = \sum_{n=1}^{N} q_{nc} u_{nc} + q_c. \tag{3.30}$$

$$\mu = \frac{C_\circ - \sum_{c=1}^{C} \frac{q_c}{\omega_c}}{\sum_{c=1}^{C} \frac{1}{\omega_c}}. \tag{3.31}$$

Now, we rewrite (3.21) as,

$$r_c = \min\left(\max\left(0, \frac{1}{\omega_c}\mu + \frac{q_c}{\omega_c}\right), 1\right). \tag{3.32}$$

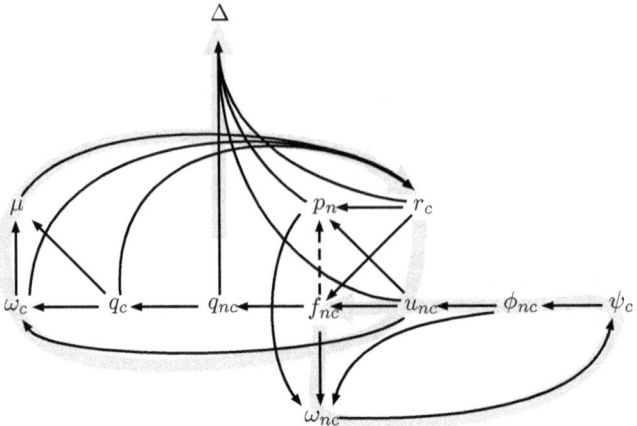

Figure 3.5: Implementation of the flow exhibited in Figure 3.4.

Moreover, we rewrite (3.18) as,

$$\Delta = \sum_{n=1}^{N} \sum_{c=1}^{C} q_{nc} \left(r_c^2 u_{nc} + (1 - r_c)^2 U \right) + \sum_{n=1}^{N} \omega_n (1 - p_n)^2 \frac{U}{2C}, \tag{3.33}$$

and (3.24) as,

$$\omega_{nc} = \frac{q_{nc}}{(\phi_{nc} + \lambda)^2}. \tag{3.34}$$

From a practical perspective, the method proposed in this paper is implemented as the Matlab class *Selma*, as shown in Figure 3.1. In Chapter 4 we provide multiple examples for the execution of the developed method within the context of different problem classes. There, for each problem class, a child class carries out tasks such as loading the input data and also generating problem class-specific output visualizations. Additionally, the child classes define the data item- and model-specific functions $\Psi_\circ(\cdot)$, $\Psi(\cdot)$ and $\phi(\cdot)$ and set the value of φ. The core operations of the algorithm, however, i.e. the process developed in Section 3.5 and this section, are maintained in the parent class. Hence, as also stated before, a single implementation of the developed algorithm is utilized for effectively any problem class, given that the mathematical entities mentioned above are provided.

Algorithm 1 shows the steps involved in the implementation of the developed algorithm. This process is pictorially depicted in Figure 3.5 as well. As seen in this figure, the implementation of the developed method contains two interlocked loops and a common link to the value of the objective function. Here, the first loop maintains the relationship between r_c, f_{nc} and p_n and the second loop updates the cluster representations.

Input: Weighted Set of Data Items **X**, Estimated number of clusters C_o
Output: Cluster Representations $\psi_1, \cdots, \psi_C$, Classification Results c_n
$C = \kappa C_o$;
Call $\Psi_o(\mathbf{X})$ in order to produce $\psi_1, \cdots, \psi_C$;
for $c = 1$ to C do Set $r_c = 1$;
/* Alternating Optimization (Sections 3.5 and 3.9)   */
while *True* do
    for $n = 1$ to N do
        for $c = 1$ to C do Calculate ϕ_{nc} and u_{nc} using (3.6);
        for $c = 1$ to C do Calculate v_{nc} using (3.28);
        Calculate v_n using (3.28);
        for $c = 1$ to C do Calculate f_{nc} using (3.29);
        Calculate p_n using (3.29);
    end
    for $c = 1$ to C do
        for $n = 1$ to N do Calculate q_{nc} using (3.30);
        Calculate q_c using (3.30);
        Calculate ω_c using (3.30);
    end
    Calculate μ using (3.31);
    for $c = 1$ to C do
        Calculate r_c using (3.32);
        for $n = 1$ to N do Calculate ω_{nc} using (3.34);
    end
    Calculate Δ using (3.33);
    /* Cluster Maintenance (Section 3.7)   */
    if *significant change in Δ is not registered in multiple iterations* then
        if $C \leq 1$ then Break;
        if $\min\{r_c\} < \frac{1}{2}$ then Delete the corresponding cluster;
        else Break;
    end
    if *No cluster was deleted* then
        for $c = 1$ to C do Calculate ψ_c using (3.23);
    end
end
/* Classification and Outlier Detection (Section 3.6)   */
for $n = 1$ to N do
    if $p_n < \frac{1}{2}$ then $c_n = 0$;
    else $c_n = \arg_c \max\{f_{nc}\}$;
end

Algorithm 1: Outline of the algorithm developed in this paper.

Chapter 4

Experimental Results

4.1 Algorithm Overview

Figure 4.1 shows an overview of the developed algorithm from an input/output perspective. In this experiment, the developed algorithm is applied on a 2dl problem instance which contains the 1,588 data items shown in Figure 4.1(a). Here, the sizes of the data items denote their value of ω_n. The reader is referred to Appendix A for the definition of 2dl and other problem classes. Here $C_\circ = 4$ and the algorithm starts with 8 clusters. After 116 iterations and 342 milliseconds of processing, the developed algorithm converges to the 4 clusters shown in Figure 4.1(b). As a result, the input set of data items is classified, as seen in Figure 4.1(c).

Figures 4.1 (a)-(c) show the inputs and outputs of the developed algorithm. These figures are augmented with Figures 4.1 (d)-(g) which denote the internal state of the developed algorithm as it transitions into its final converged situation. Here, we review these figures.

Figures 4.1(d), (e), and (f) each show multiple histograms stacked together on a horizontal line in order to demonstrate the gradual modification of these sets of probabilistic identifiers. In effect, each line in either of these figures demonstrates the histogram of the corresponding set of variables at the stated iteration. Hence, for example, Figure 4.1(d) shows that the clustering algorithms starts with f_{nc} values which are asymmetrically distributed around a central value which is less than 0.5. These variables then evolve towards the histogram shown at iteration 116, i.e. the last iteration. This histogram has two visible modes, one close to one and one less than half. These two modes correspond to data items which are either assigned or not assigned to different clusters with hight probability.

Similarly, Figure 4.1(e) demonstrates the progression of p_n values as the algorithm steps towards

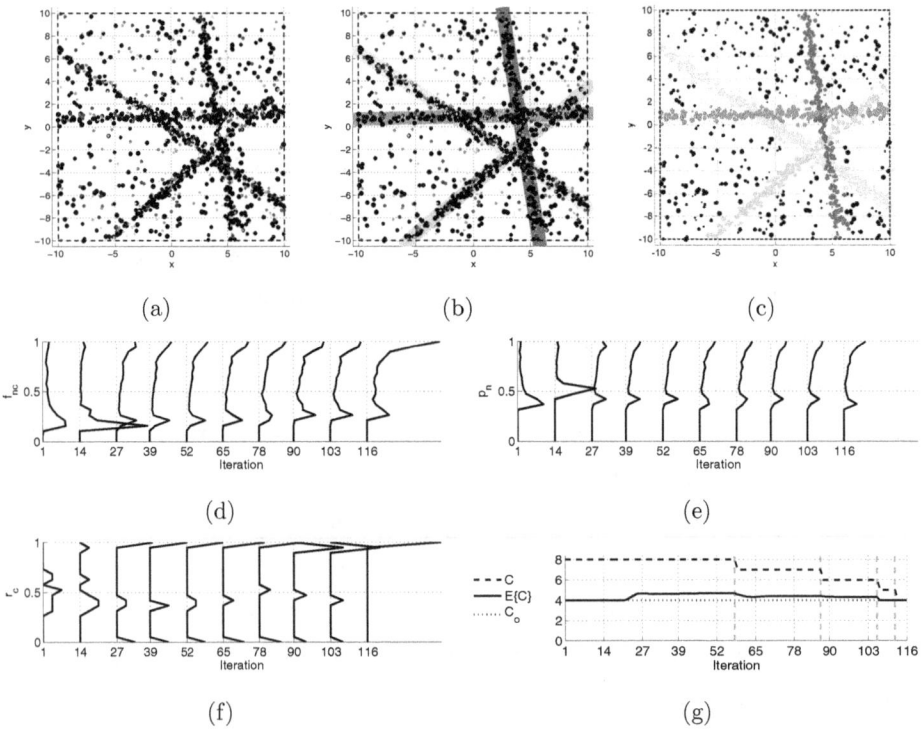

Figure 4.1: Overview of the developed algorithm from an input/output perspective for a 2dl problem instance. (a) Input set of data items. (b) Converged clusters. (c) Classification of input data items into clusters and outliers. (d), (e), (f), and (g) Internal state of the clustering algorithm. (d) Histogram of f_{nc} during the convergence of the algorithm. (e) Histogram of p_n during the convergence of the algorithm. (f) Histogram of r_c during the convergence of the algorithm. (g) Value of $C_\circ$ compared to the values of C and $E\{\mathbf{C}\}$ during the convergence of the algorithm.

convergence. Here, again, we note a relatively symmetrical p_n distribution at the beginning of the process and a histogram with two, less bold, modes above and below half after convergence. Figure 4.1(f), on the other hand, shows that the algorithm starts with r_c values throughout the $[0, 1]$ interval and travels through a bi-modal r_c distribution, with values at the two extremes. Nevertheless, we observe that the values close to zero disappear gradually. This is in fact the process of cluster maintenance in execution. Finally, as the proposed method converges, there are only $r_c \simeq 1$ values in the system. This is evident in the steep peak in the r_c histogram at iteration 116.

Finally, Figure 4.1(g) shows the changes in the values of $E\{\mathbf{C}\}$ and C as the process moves towards convergence. Here, the vertical dashed lines indicate iterations at which the cluster maintenance process is activated. These steps are, by definition, the times at which C decrements by one. Moreover, the horizontal dotted line in Figure 4.1(g) shows the value of $C_\circ$. We note that, as expected, C starts at $\kappa C_\circ$. Additionally, we note that $E\{\mathbf{C}\} = C_\circ$ is only enforced *loosely* during the process, as visibly seen in Figure 4.1(g), in which the $E\{\mathbf{C}\}$ and $C_\circ$ curves do not necessarily collide at every iteration. In fact, we notice that it appears that $E\{\mathbf{C}\}$ is always greater than or equal to $C_\circ$. This observation may result in the conjecture that r_c is always adjusted from the bottom in (3.22), i.e. a negative value of r_c is always pushed up to zero. In fact, however, we have registered both instances of $r_c^\star < 0$ and $r_c^\star > 1$. We argue that, any claim regarding the status of r_c prior to (3.22) must review the optimal point of (3.18) relative to r_c. We note that Δ is a quadratic function of r_c and that it is intersected with a plane. We also note that, for this experiment, at convergence we have $C = C_\circ = E\{\mathbf{C}\}$. This equilibrium may not in fact be the case for all settings. One may investigate the satisfaction of this equilibrium as an evidence that $C_\circ$ has been selected correctly. Further analysis of these and other topics related to $C_\circ$ are outside the scope of this paper.

4.2 φ Diagnostics

As noted in Section 3.2, the behavior of the developed method is regulated by the two parameters λ and U. Moreover, in Section 3.8, we provided two deterministic construction processes which generate the optimal values of these parameters in the context of any arbitrary problem class. That process assigned appropriate values to the two parameters φ_0 and φ_1 and stated the relationship between these parameters and λ and U. We also discussed that we recommend that φ_0 and φ_1 are

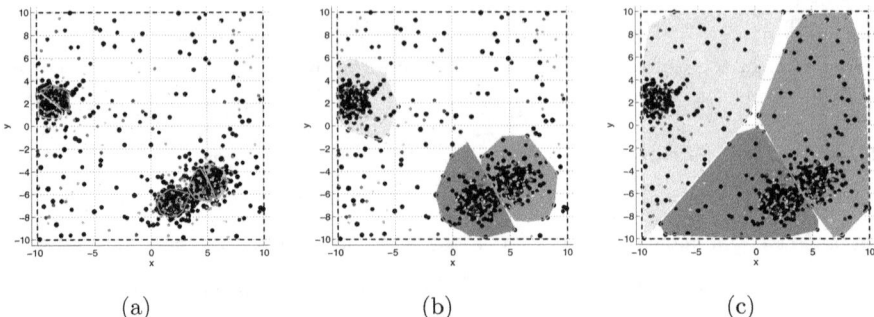

Figure 4.2: Assessment of the impact of the value of φ on the results of the developed algorithm for the 2de problem class. (a) $\varphi = \frac{1}{10}\varphi_{2de}$. (b) $\varphi = \varphi_{2de}$. (c) $\varphi = 10\varphi_{2de}$. Here, φ_{2de} denotes the optimal value of φ for the 2de problem class, as discussed in Appendix A.2.

assigned to a unique value, which we denoted as φ^2.

We argue that the proper setting of the *scale* parameter is one of the most critical aspects which govern the performance of any clustering algorithm. In essence, scale dictates *how far a data item can be from a cluster and still belong to it*. From the opposite perspective, too, the value of the scale parameter governs the sizes of the clusters. We note that in the present work, φ is the single configuration variable which controls the scale at which the developed algorithm functions. Hence, here, we review the impact of φ on the output of the developed algorithm and provide some insights.

Figure 4.2 shows the results of executing the developed method on a 2de problem instance. In fact, the three results exhibited in this figure correspond to three different values of φ. We pick these values in accordance with the optimal φ value corresponding to the 2de problem class, as given in Appendix A.2. In effect, Figure 4.2 addresses a case in which φ is underestimated and Figure 4.2(c) provides a case in which φ is over-estimated. Figure 4.2(b), on the other hand, utilizes $\varphi = \varphi_{2de}$.

The size of the dots in Figure 4.2 denotes the value of ω_n and the colored polygons identify the convex hull of the data items which are classified into the same cluster.

The two sets of results given in Figures 4.2(a) and (c) correspond to the two extreme cases in which φ is an order of magnitude different from φ_{2de}. Hence, as expected, these two scale settings result in clusters which are significantly smaller or larger than expected. In fact, in Figure 4.2(a), each cluster is decomposed into multiple clusters and in Figure 4.2(c) each cluster has extended its territory to a significantly large set of data items which reside outside of its concentrated core.

We note that the three sets of results provided in Figure 4.2 all correspond to the same value of $C_\circ$, i.e. $C_\circ = 3$. This setting, in the case of Figure 4.2(c), results in three clusters with r_c values which each exceed $1 - 10^{-3}$ (in this case $E\{\mathbf{C}\} = 2.999$). In the case of Figure 4.2(a), however, the developed algorithm generates five clusters, which in fact belong to three sets with very similar ψ_c vectors. In other words, two pairs of clusters in Figure 4.2(a) each generate a full circle. The first pair is $(\psi 1, \psi 2)$ and the second pair is $(\psi 3, \psi 5)$. Here, $\psi 1 = [-8.559, 2.069] \simeq [-8.561, 2.066] = \psi 2$ and $\psi 3 = [5.201, -4.982] \simeq [5.201, -4.982] = \psi 5$. We note that in this case, $|E\{\mathbf{C}\} - 3|$ is less than 10^{-8}. In other words, we observe that when φ is too small, clusters replicate each other and it is possible that multiple clusters may converge to the same cluster representation while they claim different subsets of the input data items. Hence, when the developed algorithm generates many small clusters, it may be an indication that φ should be increased. On the other hand, when the number of clusters is acceptable but they contain too many outliers, we advise that φ should be reduced. We note that the three sets of results provided in Figure 4.2 all correspond to the same value of $C_\circ$, i.e. $C_\circ = 3$. This setting, in the case of Figure 4.2(c), results in three clusters with r_c values which each exceed $1 - 10^{-3}$ (in this case $E\{\mathbf{C}\} = 2.999$). In the case of Figure 4.2(a), however, the developed algorithm generates five clusters, which in fact belong to three sets with very similar ψ_c vectors. In other words, two pairs of clusters in Figure 4.2(a) each generate a full circle. The first pair is $(\psi 1, \psi 2)$ and the second pair is $(\psi 3, \psi 5)$. Here, $\psi 1 = [-8.559, 2.069] \simeq [-8.561, 2.066] = \psi 2$ and $\psi 3 = [5.201, -4.982] \simeq [5.201, -4.982] = \psi 5$. We note that in this case, $|E\{\mathbf{C}\} - 3|$ is less than 10^{-8}. In other words, we observe that when φ is too small, clusters replicate each other and it is possible that multiple clusters may converge to the same cluster representation while they claim different subsets of the input data items. Hence, when the developed algorithm generates many small clusters, it may be an indication that φ should be increased. On the other hand, when the number of clusters is acceptable but they contain too many outliers, we advise that φ should be reduced.

Figure 4.3 shows the results of a similar experiment carried out on a 2dl problem instance. Here, in line with the results shown in Figure 4.2, the results of setting φ to the three values of $\frac{1}{10}\varphi_{2dl}$, φ_{2dl}, and $10\varphi_{2dl}$ are exhibited.

The result carried in Figure 4.3(a) corresponds to the situation in which φ is significantly smaller than the value recommended based on the properties of the underlying physical model. In this case, we observe that the algorithm does not "notice" the sections of the data which contain the clusters.

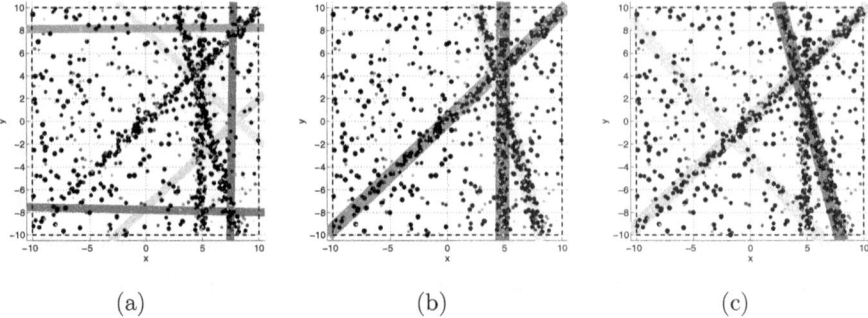

(a) (b) (c)

Figure 4.3: Assessment of the impact of the value of φ on the results of the developed algorithm for the 2dl problem class. (a) $\varphi = \frac{1}{10}\varphi_{2dl}$. (b) $\varphi = \varphi_{2dl}$. (c) $\varphi = 10\varphi_{2dl}$. Here, φ_{2dl} denotes the optimal value of φ for the 2dl problem class, as discussed in Appendix A.3.

In other words, small values of φ can force the algorithm to not be able to detect concentrations of the data items, as they "perceive" the data items to be equally *too far* and *insignificant* and thus miss the fact that a subset of the data items forms a cluster.

The case of an inflated φ value, as exhibited in Figure 4.3(c), shows the opposite situation, in which a cluster repels other clusters from a formation of data items which do not belong to the cluster, but, nevertheless, are "perceived" too close to another cluster in order to be eligible to be considered as a separate cluster.

On the contrary to both Figures 4.3(a) and (c), the result shown in Figure 4.3(b) denotes a φ value which matches what the theory predicts to be the optimal value. For this result, redundant clusters are correctly removed and the disjoint clusters are correctly identified.

4.3 Impact of κ

As discussed in Section 3.3, this paper utilizes the concept of cluster fuzziness and couples it with the notion of the *ideal number of clusters*, as denotes by C_o. In other words, the method developed in this paper allows for an excessive number of clusters to compete for the data items and inherently provides a means of cluster pruning. This is mathematically modeled through the *cluster redundancy factor*, κ. Nevertheless, it is a valid question whether the particular value of κ utilized in one experiment may have a significant impact on the outcome of the process. If such dependency is not noticed, then one may suggest that, for practical purposes, κ acts as a cost-

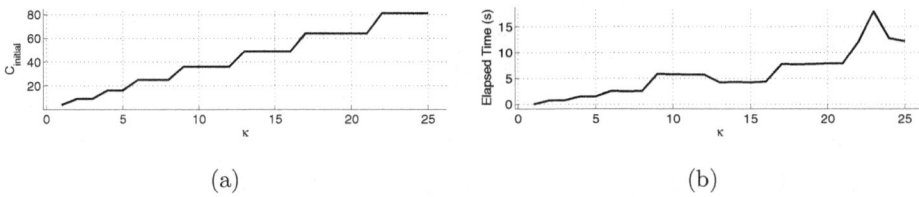

Figure 4.4: Summary of the execution results for a 2dc problem instance with three clusters and different κ values. (a) Number of clusters initialized at the beginning of the process. (b) Total elapsed time.

multiplier which can increase the possibility of *desirable* clustering. To be phrased differently, one may increase κ as long as resource budgets allow and reasonably expect better clustering results.

Figure 4.4 shows summary of the execution results corresponding to a unique 2dc problem instance with different values of κ. There are three clusters in the set of data items utilized in this problem instance. Here, Figure 4.4(a) shows the value of C at the beginning of the process. Note the staircase relationship between $C_{initial}$ and κ. This is due to the fact that, in this problem class, $\kappa C_\circ$ is updated to $\lceil \sqrt{\kappa C_\circ} \rceil^2$. For this range of κ values, which expands from 1 to 25, the developed method always converges to 3 clusters. Figure 4.4(b) exhibits the total elapsed time for these processes.

Figure 4.5 shows some of the utilized κ values. Here, the top row indicates the clusters as they are initialized and the bottom row exhibits the output of the clustering procedure. We observe that independent of the value of κ, the developed process converges to the same number of cluster. The converged clusters, too, are similar, upon visual inspection.

4.4 Comparative Assessment

It is a fair argument that the usefulness of a novel clustering algorithm is not fully known until it has been rigorously examined against previous approaches available in the literature. This argument, however, must be considered carefully and in the context.

From an epistemological perspective, the fact that a particular set of experiments shows that algorithm A is superior to algorithm B is not a justification for an overarching argument regarding the performance of the two algorithms. In essence, one may, most probably unintentionally, produce a set of experiments for which algorithm A behaves more favorably than algorithm B. The fact

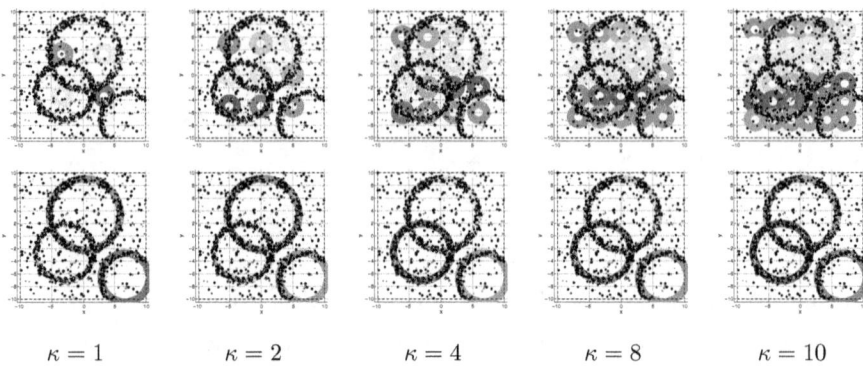

Figure 4.5: Initial and converged clusters for different κ values. Top) Clusters as they are initialized. Bottom) Converged clusters.

that for many fuzzy clustering algorithms there are one or more configuration parameters which greatly impact the output of the algorithm, makes the situation extremely more complex. In such circumstances, the question really is not "*how well does algorithm A function for problem instance P*", but instead, one is faced with the more critical question of "*Does algorithm A function acceptably for problem instance P if no human consciousness provides assistance?*" In other words, it does not matter too much if an algorithm *can* produce acceptable results for a number of problem instances. Instead, one must ask if lack of human assistance may jeopardize the performance of a fuzzy clustering algorithm. In the presence of non-managed parameters, the answer to this questions seems to be positive. Here, a non-managed parameter is one for which no deterministic construction mechanism is proposed and that the task of setting the parameter is left for a user to fulfill. Hence, we argue that any fuzzy clustering algorithm which involves a non-managed parameter, for example a regularization coefficient which is to be set by the user, inherently requires user-assistance and is therefore not par for comparison with the algorithm developed in this paper. From an epistemological perspective, the fact that a particular set of experiments shows that algorithm A is superior to algorithm B is not a justification for an overarching argument regarding the performance of the two algorithms. In essence, one may, most probably unintentionally, produce a set of experiments for which algorithm A behaves more favorably than algorithm B. The fact that for many fuzzy clustering algorithms there are one or more configuration parameters which greatly impact the output of the algorithm, makes the situation extremely more complex. In such circumstances, the question really is not "*how well does algorithm A function for problem instance P"*, but instead, one is faced with the more critical question of *"Does algorithm A function acceptably for problem instance P if no human consciousness provides assistance?"* In other words, it does not matter too much if an algorithm *can* produce acceptable results for a number of problem instances. Instead, one must ask if lack of human assistance may jeopardize the performance of a fuzzy clustering algorithm. In the presence of non-managed parameters, the answer to this questions seems to be positive. Here, a non-managed parameter is one for which no deterministic construction mechanism is proposed and that the task of setting the parameter is left for a user to fulfill. Hence, we argue that any fuzzy clustering algorithm which involves a non-managed parameter, for example a regularization coefficient which is to be set by the user, inherently requires user-assistance and is therefore not par for comparison with the algorithm developed in this paper.

Additionally, a validation framework which is solely reliant on the outputs of the algorithms, lacks conviction regarding the construction of the corresponding approaches. In other words, even if an algorithm is shown to *work well*, based on experimental results, as long as the mathematical framework behind the model is not justified, one may attribute the supposed efficiency of the corresponding algorithm to the intuition and luck of the researchers. We argue that such notions are useful but not objectively reliable and, that, therefore, validation must start at the level of mathematical modeling and formalization.

Nevertheless, it is extremely useful to perform comparative assessment of a novel algorithm relative to previous works in the literature. For this purpose, we compare the algorithm developed in this paper with FCM. We choose FCM, because it is the grand work which has bloomed into a majority of other works in the field of fuzzy clustering. Additionally, FCM does not depend on any non-managed parameters. It must be emphasized, however, that FCM is an Euclidean distance algorithm, which has also been extended to the Mahalanobis and a few other distance functions. Nevertheless, FCM is not historically intended to be a class-independent fuzzy clustering algorithm. Hence, we generalize FCM using the mathematical framework developed in this work and therefore address FCM in the context of different problem classes.

Here, we carry pairs of results generated by FCM and the developed method for problem instances in different problem classes. We emphasize that the circumstances governing the execution of the two algorithms are identical, to the extent possible due to the intrinsic differences of the algorithms, and that neither of the two algorithms receives any user assistance.

Figures 4.6, 4.7, and 4.8 carry comparative experimental results generated by the proposed method and FCM. Here, the results are shown side-by-side in order to allow direct comparison.

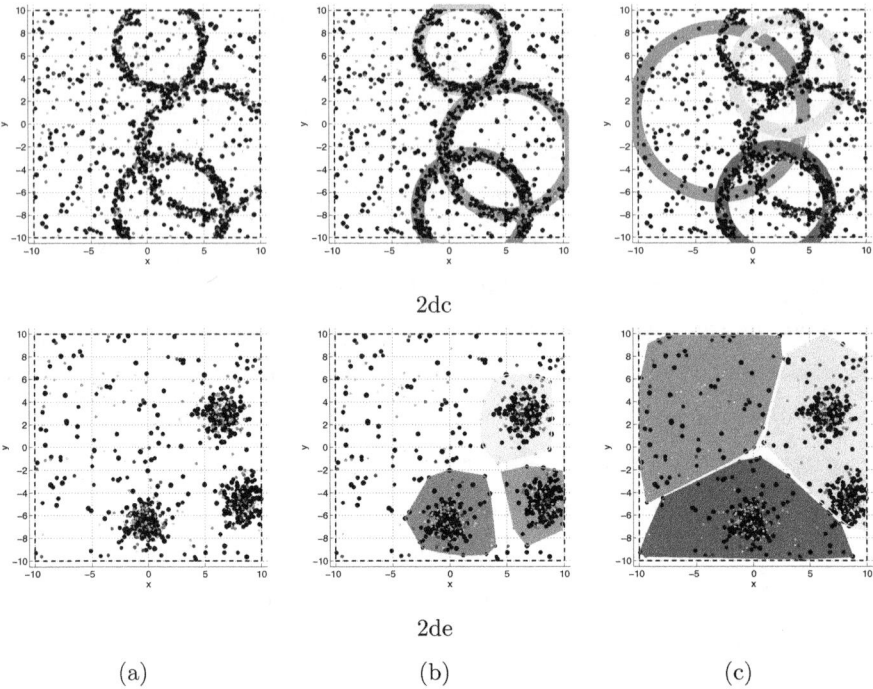

Figure 4.6: Pairs of results generated by the proposed method and FCM, carried side-by-side in order to allow comparison. (a) Input data items. (b) Output of the proposed method. (c) Output of FCM.

Note that in this set of experiments, we have chosen problem instances for most of which $C_\circ = 3$ in order to facilitate visualization and comparison of the results. Table 4.1 carries the numerical values corresponding to these experiments.

Top row in Figure 4.6 shows comparative results for a problem instance in the 2dc problem class. We observe that while the proposed algorithm converges to the three circular clouds of points in the input data, two of the FCM clusters converge to sections in the data which do not constitute the geometrical structure which is conveyed by the notion of homogeneity identified by the problem class. We also note that in the results of FCM, the third cluster indeed converges to the vicinity of one of the clusters present in the data. Nevertheless, we observe evident misalignment between the cluster representation and the data. We argue that both observed phenomena, i.e. convergence of clusters to irrelevant sections of data, as well as cluster-data misalignment, are direct results of the fact that FCM does not contain a robustness component, and in fact does not intend to attempt robustness.

The case shown in the second row in Figure 4.6 corresponds to the classical 2de problem class. This problem class is in fact commonly considered as *the* problem class for data clustering. Nevertheless, we notice two important properties in the results generated by FCM. First, FCM clusters span to cover the entire set of data items. This is in fact an inherent property of FCM, in which all data items are assumed to be inliers. In other words, the FCM question is *"Which cluster better represents this data item?"*. The proposed method, on the other hand, augments that question with an *"if any"* clause. Additionally, we note that in the results of FCM, two clusters are lumped into one and one cluster converges to a section of the data which corresponds to outliers. Neither of the above unwanted phenomena is observed in the output of the proposed method. In fact, the clusters generated by the proposed method are compact, i.e. they do not bloat into the boundaries of the input data, and they in fact correspond to clusters in the data.

Top row in Figure 4.7 shows sample results corresponding to the 2dl problem class. This problem class is, loosely speaking, the linear equivalent of the 2dc problem class, which is exhibited in the top row of Figure 4.6. Similar to the observations made in that figure, we notice that FCM clusters are highly prone to both converging to outlier sections of the data as well as being misaligned to the inherent clusters in the data.

Bottom row in Figure 4.7 is similar to the 2dl problem class as well. Here, the intent is to cluster 3D points into planes. The data utilized in this experiment corresponds to a corner of a room in which three human figures are present. We notice that the proposed method is capable of converging to the three walls in the scene. FCM, however, generates clusters which are "trapped" by the outlier sections of the data, here anything in the scene which does not correspond to the walls. Additionally, we note that the human figures are painted in black in the results of the proposed method. This indicates that the proposed method has been successfully capable of recognizing that these sections of the data are outliers. In the results of FCM, however, the human figures are amended to the converged clusters.

Finally, Figure 4.8 compares the performance of the developed method with that of FCM for the two image-domain problem classes ics and ighe. While meaningful comparison is less feasible in the image domain, the histogram-domain representation given on the third row in Figure 4.8 provides informative insights into the performance of the two algorithms. These insights are in fact in line with the observations made in the preceding paragraphs regarding the differences between the performance of the two algorithms for other problem classes.

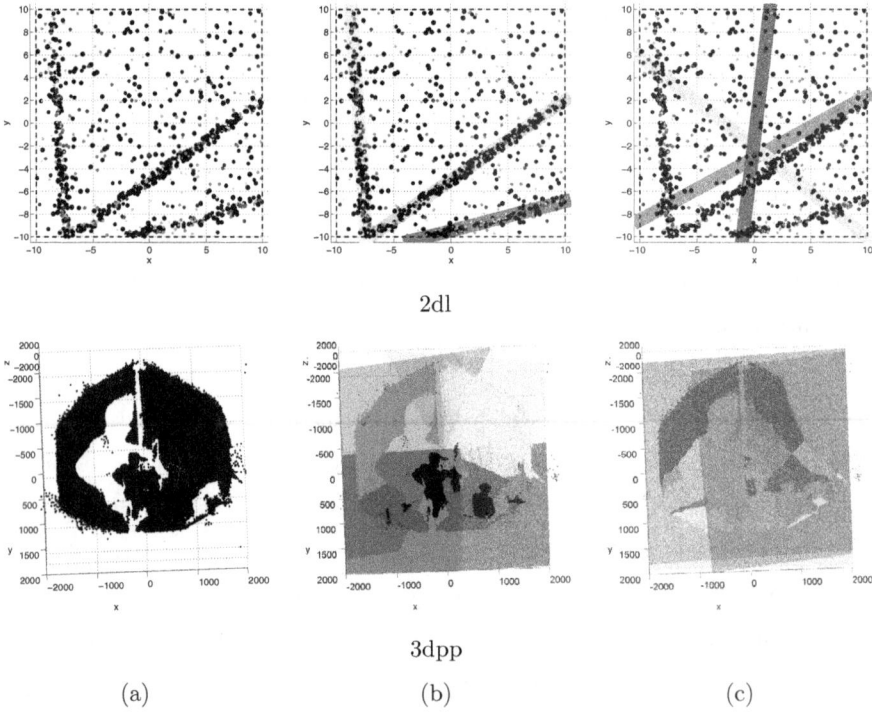

Figure 4.7: Pairs of results generated by the proposed method and FCM, carried side-by-side in order to allow comparison. (a) Input data items. (b) Output of the proposed method. (c) Output of FCM. The input data used in 3dpp experiments is courtesy of *Epson Edge, Epson Canada Limited*.

43

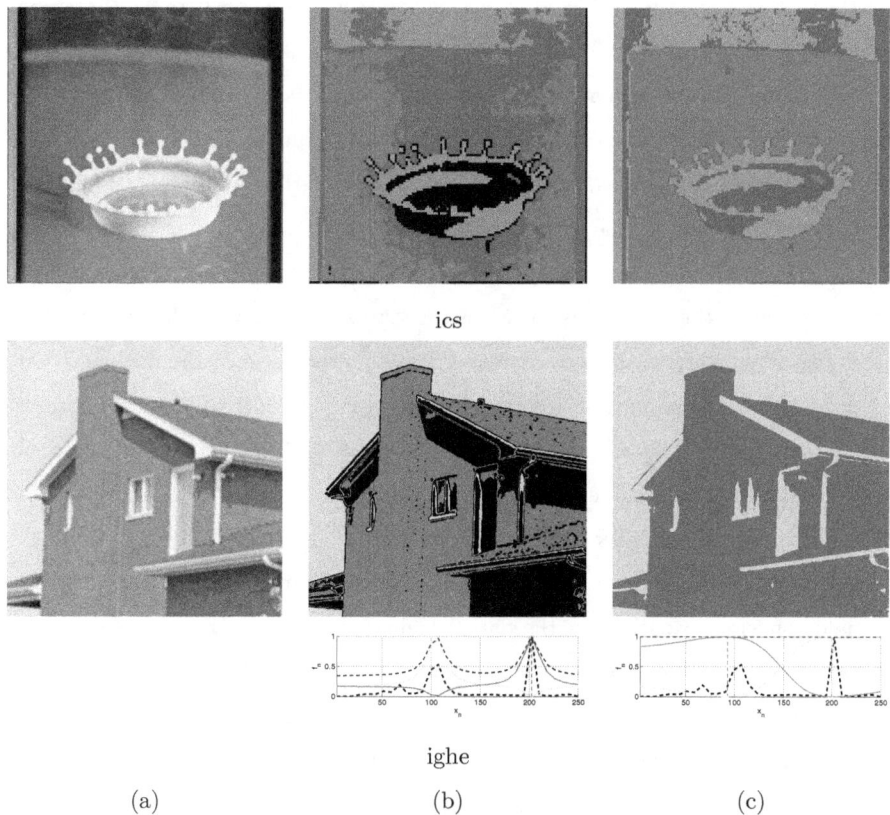

Figure 4.8: Pairs of results generated by the proposed method and FCM, carried side-by-side in order to allow comparison. (a) Input data items. (b) Output of the proposed method. (c) Output of FCM.

In the histogram-domain representation of the clustering results, as carried in the third row of Figure 4.8, the thick dashed line indicates the histogram of the input data items. The thin dashed line, on the other hand, denotes the values of p_n. As expected, in FCM we have $p_n \equiv 1$, and, therefore, the thick dashed line is always at unity. This property of FCM, and many other fuzzy clustering algorithms, can be observed in the image-domain representation as well, where c_n is non-zero for every x_n.

The colored lines in the third row of Figure 4.8 identify the clusters. Through comparing these curves for the proposed method and FCM we learn about the differences between the two methods. For example, FCM tends to assign high membership values to data items when only one cluster is contending for them. This phenomenon is visible, for example, in the elevation of the membership curve corresponding to the second cluster at the right edge of the histogram in Figure 4.8(c). The proposed method, however, reduces the interest in data items as they move far from any particular cluster, although there may not be a contending cluster. In other words, the "power dynamics" in FCM is between contending clusters while in the proposed method, every cluster must compete against the tendency of the data items to accept the status of being an outlier as well.

Additionally, it is informative to observe the misalignment between the first cluster from the right in Figure 4.8(c). In comparison, the same cluster in Figure 4.8(b) aligns more symmetrically with the same section of the data items. We conjecture that lack of a robustification mechanism in FCM allows far data items to influence a cluster, and, therefore, pull it out of alignment with the set of data items which have higher membership values to it.

The comparative results carried in Figures 4.6, 4.7, and 4.8 indicate that it is likely that FCM clusters would converge to outliers in the data. Additionally, it is commonly observed that one FCM cluster would converge to the concatenation of two clusters in the data. When a cluster correctly converges to the data too, it is observed that evident misalignment between the cluster and the data is a highly probable situation in FCM results. The proposed method, on the other hand, is observed to be capable of avoiding these pitfalls.

Review of the numerical results carried in Table 4.1 shows that the developed method is always more expensive than FCM. In fact, 3dpp is the problem class in which the time elapsed by the two algorithms is the closest. Even for this problem class, the proposed method is 1.93 times more expensive than FCM. In fact, in average, the proposed method requires 3.12 times more computational resources than FCM. We note that, nevertheless, the computational cost of the proposed method is within one order or magnitude of FCM.

Table 4.1: Numerical values corresponding to the comparative experimental results carried in Section 4.4.

Problem Class	N	Algorithm	$C_\circ$	C	Elapsed Time	Ratio
2dc	1,642	FCM	3	3	104 ms	7.41
		Proposed	9	3	721 ms	
2de	987	FCM	3	3	49 ms	2.93
		Proposed	9	3	144 ms	
2dl	1,166	FCM	3	3	73 ms	3.3
		Proposed	8	3	241 ms	
3dpp	42,754	FCM	3	3	1,801 ms	1.93
		Proposed	6	3	3,487 ms	
ics	16,384	FCM	3	3	119 ms	9.38
		Proposed	6	3	1,102 ms	
ighe	32	FCM	2	2	12 ms	4.61
		Proposed	4	2	44 ms	
Overall						3.12

Chapter 5

Conclusions

In this paper, we addressed the generic class-independent problem of clustering a set of data items of an arbitrary mathematical model into a set of homogenous clusters and a set of outliers. The model developed in this paper allows for the adoption of any arbitrary cluster model, as long as a set of functions which operate on data items and clusters are given as black-boxes.

We employed a purely derivation-based approach and utilized Bayesian inference in order to model the loss associated with the generic fuzzy clustering problem studied in this paper. We noted that in the past decades the research community has benefited from capitalizing on a fuzzy relationship between the data items and the clusters and also that fyzzy data items have been employed. Nevertheless, the clusters have not been transformed into fuzzy objects. We showed that this transformation is possible within the loss modeling framework developed in this paper and that this transformation has important benefits. As a result, we developed a class-independent fuzzy clustering algorithm in this paper. This algorithm is constructed from bottom to top based on clear mathematical derivation and use of metaphors and heuristics is decidedly avoided in the process.

We demonstrated the adoption of the developed algorithm for six problem classes which concern linear and quadratic cluster models in 2- and 3D spaces. We showed that within the same problem class, the developed algorithm does not need to be adjusted in order to operate on different problem instances. For different problem classes, too, although there are configuration variables which need to be set, we demonstrated that objective and deterministic processes for setting these variables exist.

To the best of our knowledge, the introduction of cluster fuzziness is an important contribution of this work. Additionally, the independence of the developed algorithm from configuration variables which are to be adjusted empirically and through trial-and-error is an important practical contribution of the present work. On the epistemological level, too, we argue that the construction mechanism developed and utilized in this paper is superior to intuition and heuristic-based approaches in the literature.

Acknowledgments

We wish to thank the management of *Intellijoint Surgical Inc.* for their support. We thank the management of *Epson Edge, Epson Canada Limited* for allowing us to use the Kinect data used in the experiments carried in this paper. This research and the writing of this paper were carried out in different cafes in the Kitchener-Waterloo region, especially in *Cafe 1842*.

Appendix A

Problem Classes

Here, the problem classes utilized in this paper are defined. For each problem class, first the mathematical models for the data items and the clusters are provided. Then, the procedure utilized in order to generate a set of data items, i.e. $\mathbf{X}$, is described. This process may potentially input the number of desired clusters, which we denote as C. This case is applicable to problem classes which utilize artificially generated data. In these cases we use,

$$\mathbf{X} = \left[\bigcup_{c=1}^{C} \mathbf{X}_c \bigcup \mathbf{X}_\circ\right] \bigcap [-R, R]^2. \tag{A.1}$$

Here, $\mathbf{X}_c$ is a sample set of $N_\circ$ data items which are generated using a unique cluster model. The set $\mathbf{X}_\circ$, on the other hand, is a sample set of $N_\circ$ data items and models background noise. The parameter R denotes the range of data and in the experiments carried in this paper we always use $R = 10$.

In order to facilitate the mathematical formulations, we utilize the two notations $\mathfrak{x}_c$ and $\mathfrak{x}_\circ$. Here, $\mathfrak{x}_c$ is one data item which is generated according to the cluster model for ψ_c and $\mathfrak{x}_\circ$ is one data item which is generated based on the model for background noise. Hence, unless specified differently, $\mathbf{X}_c$ is a set of $N_\circ$ independent realizations of $\mathfrak{x}_c$ and $\mathbf{X}_\circ$ is a set of $N_\circ$ independent realizations of $\mathfrak{x}_\circ$. Note that, unless specified differently, $\mathfrak{x}_\circ = \mathfrak{u}_2 R$. In this notation, $\mathfrak{u}$ denotes one realization of the uniform random variable defined in $[-1, 1]$ and $\mathfrak{n}$ denotes one realization of the Gaussian random variable with mean of zero and standard deviation of 1. We use prefixes in order to denote i.i.d. random variables formed into multidimensional vectors. Hence, for example, $\mathfrak{u}_2 = [\mathfrak{u}, \mathfrak{u}]^T$.

Note that, $\omega_n = \mathfrak{u}$ unless specified differently and the parameter ε in the models described here controls noise amplitude. The reader is referred to the explanations given in the following sections for the details of each model. Table A.1 carries the properties of the different problem classes utilized in this paper.

We note that, the process contained in $\Psi_\circ(\cdot)$ may potentially increase the value of C. For example, in the case of the 2dl problem class, as discussed in Appendix A.3, in fact $4\lceil\frac{1}{4}C\rceil \geq C$ clusters are generated. We tolerate this increase in the number of clusters, due to the fact that the algorithm developed in this paper is capable of handling $C > C_\circ$. When the same framework is executed in FCM-mode, however, we trim the ψ matrix in order to enforce that it satisfies $C = C_\circ$.

The single control parameter in the method developed in this paper is φ. As discussed in Section 3.8, this parameter governs the scale of the developed algorithm. Due to the significance of this parameter, we address it explicitly for every problem class which is discussed in this section. It is important to recognize, however, that the first three problem classes listed in this section utilize artificially generated data while the last three utilize data items which correspond to some real world phenomenon. Hence, we provide estimates of φ for artificially generated data and justify the choice of φ for real world data.

Figure A.1 carries the geometrical models associated with the problem classes utilized in this paper. These figures describe the relationship between the data items and the clusters as well.

53

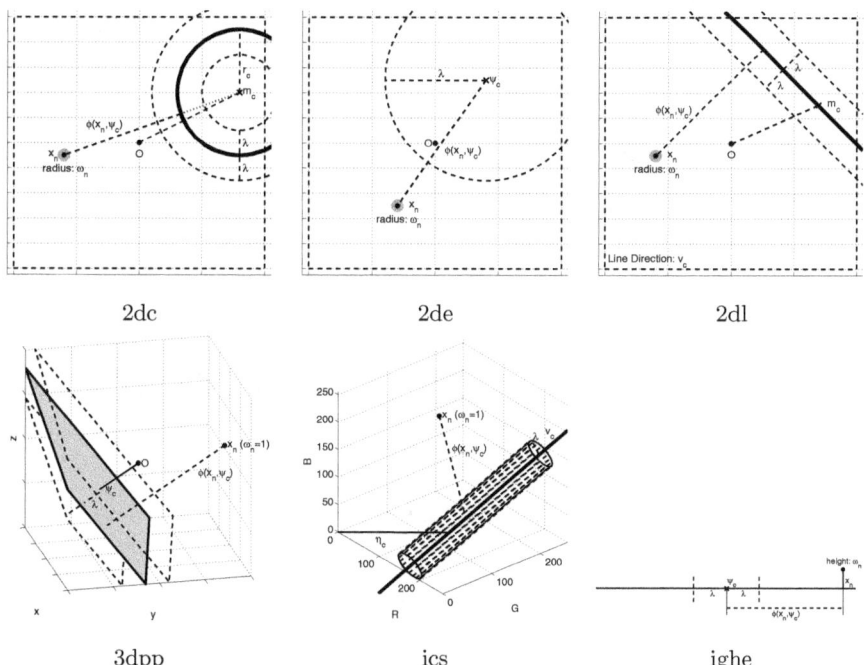

Figure A.1: Geometrical models corresponding to the problem classes utilized in this paper.

Table A.1: Properties of the problem classes utilized in this paper. Cells denoted as N/A identify aspects which are not applicable to the corresponding problem classes.

	x_n	ψ_c	$\mathfrak{x}_c$	ε	φ	N_o	$\phi(x_n, \psi_c)$
2dc	$x_n \in \mathbb{R}^2$	$\psi_c = [m_c, \rho_c]$ $m_c \in \mathbb{R}^2$ $\rho_c > 0$	$\mathfrak{x}_c = m_c + \rho_c(1 + \mathfrak{n}\varepsilon)\angle 2\pi\mathfrak{u}$ $m_c = \mathfrak{u}_2 R$ $\rho_c = \frac{1}{2}R + \mathfrak{n}$	0.2	$\varepsilon(1+\varepsilon)(R+4)^2$	500	$\left[\|x_n - m_c\|^2 - \rho_c^2\right]^2$
2de	$x_n \in \mathbb{R}^2$	$\psi_c \in \mathbb{R}^2$	$\mathfrak{x}_c = \psi_c + \mathfrak{n}_2 R\varepsilon$ $\psi_c = \mathfrak{u}_2 R$	0.1	$2R\varepsilon\sqrt{10}$	250	$\|x_n - \psi_c\|^2$
2dl	$x_n \in \mathbb{R}^2$	$\psi_c = [m_c, v_c]$ $m_c \in \mathbb{R}^2$ $v_c \in \mathbb{R}^2$ $\|v_c\| = 1$	$\mathfrak{x}_c = m_c + \mathfrak{n}R\sqrt{2}v_c + \mathfrak{n}_2 R\varepsilon$ $m_c = \mathfrak{u}_2 R$ $v_c = 1\angle 2\pi\mathfrak{u}$	0.025	$2R\varepsilon\sqrt{10}$	500	$\|x_n - m_c - v_c^T(x_n - m_c)v_c\|^2$
3dpp	$x_n \in \mathbb{R}^3$	$\psi_c \in \mathbb{R}^3$	N/A	N/A	$200\ mm$	N/A	$\frac{1}{\|\psi_c\|^2}\left(\psi_c^T x_n - \|\psi_c\|^2\right)^2$
ics	$x_n \in \mathbb{R}^3$	$\psi_c = [m_c, v_c]$ $m_c \in \mathbb{R}^3$ $v_c \in \mathbb{R}^3$ $\|v_c\| = 1$	N/A	N/A	$\frac{1}{10}255\sqrt{3}$	N/A	$\|x_n - m_c - v_c^T(x_n - m_c)v_c\|^2$
ighe	$x_n \in \mathbb{R}$	$\psi_c \in \mathbb{R}$	N/A	N/A	$\frac{1}{10}255$	N/A	$(x_n - \psi_c)^2$

A.1 Circle-Finding in 2D Data (2dc)

This problem class is concerned with finding circular clusters, which are represented with a center and a radius, in 2D Data. The input set of data items in this problem class is generated through the superposition of C subsets of points, each one of which corresponds to randomly selected samples on the circumference of a circle. We add Gaussian noise along the radius of the circle to the points in each subset and also include a subset of randomly selected points to $\mathbf{X}$ as well.

In this problem class, $\Psi_\circ(\cdot)$ generates C uniformly distributed circles in $[-R, R]^2$ with radius of λ. The cluster fitting function utilized for this problem class is based on the method proposed by Kasa [115]. It is known that this technique requires that a large portion of the perimeter of the target circle is available [116]. We note that more robust alternatives are available and may be applicable to the present framework. The reader is referred to surveys of circle fitting algorithms in [117, 118].

In order to estimate φ for this problem class, we note that for x_n which belongs to ψ_c we have,

$$\phi(x_n, \psi_c) = \rho_c^4 \varepsilon^2 \left(\mathfrak{n}^2 \varepsilon + 2\mathfrak{n}\right)^2. \tag{A.2}$$

Now we insert the equality $p\{\mathfrak{n} < 2\} \simeq 0.9953$, which is based on the CDF of a standard Gaussian distribution, in (A.2) and generate a not-too-large upper-bound for $\phi(x_n, \psi_c)$ as $\varphi = \varepsilon(1+\varepsilon)(R+4)^2$.

A.2 Euclidean Clustering of 2D Data (2de)

This problem class addresses the classical case of finding dense clusters in weighted 2D data. Here, dense is defined in terms of inter-cluster proximity of points. In this problem class, we generate $\mathbf{X}$ as the superposition of C subsets, each generated through adding Gaussian noise to a randomly selected point in the working range. We also add uniform background noise to $\mathbf{X}$. Additionally, in this problem class, $\Psi_\circ(\cdot)$ generates C uniformly distributed points in $[-R, R]^2$ and $\Psi(\cdot)$ calculates the weighted sample mean.

For x_n which belongs to ψ_c we write,

$$\phi(x_n, \psi_c) = R^2 \varepsilon^2 \chi_2^2. \tag{A.3}$$

Now, using the CDF for a Chi-squared random variable with two degrees of freedom, we write $p\{\chi_2^2 < 10\} = 0.9933$. Hence, we utilize the not-too-large upper-bound estimate derived through this calculation for φ and write $\varphi = 2R\varepsilon\sqrt{10}$.

A.3 Line-Finding in 2D Data (2dl)

This problem class addresses the problem of finding lines in a set of weighted 2D points. We generate the input set of data items in this problem class through superimposing C subsets of points on randomly selected lines in $\mathbb{R}^2$, where the points are contaminated by Gaussian noise. We also add background noise to $\mathbf{X}$.

In this problem class, $\Psi_\circ(\cdot)$ generates $\lceil \frac{1}{4}C \rceil$ lines at each one of the four directions of 0, $\frac{\pi}{4}$, $\frac{\pi}{2}$, and $\frac{3\pi}{4}$. These lines pass through the uniformly distributed percentiles of the marginal distributions of $\mathbf{X}$. Moreover, $\Psi(\cdot)$ calculates sample mean and then performs weighted Singular Value Decomposition (SVD) in order to produce the direction of the line corresponding to each cluster. Direct derivation shows that the estimate calculated for φ in Appendix A.2 is applicable for this problem class as well.

A.4 Plane-Finding in Range Data (3dpp)

This problem class is concerned with finding planar sections in range data. The input data in this problem class contains 3D points captured by a *Kinect 2* sensor. The depth-maps used in this experiment are captured at the resolution of 424×512 pixels. Here, intrinsic parameters of the camera are acquired through the Kinect SDK and each data item in this problem class has the weight of one, i.e. $\omega_n \equiv 1$.

We model every cluster as the smallest vector, from a Euclidean norm perspective, which connects the origin to the plane which represents that cluster. In this problem class, $\Psi_\circ(\cdot)$ generates $\lceil \frac{1}{3}C \rceil$ planes perpendicular to each one of the axes. These planes pass through $\lceil \frac{1}{3}C \rceil$ uniformly-distributed percentiles of the marginal distributions of $\mathbf{X}$. Moreover, $\Psi(\cdot)$ utilizes sample mean and weighted SVD in order to fit a cluster to a given set of weighted data.

In [119], the author utilizes the distance threshold of 20 mm in order to determine whether a point is an inlier when running RANdom SAmple Consensus (RANSAC) [120]. The data acquisition process in that work utilizes an ASUS Xtion Pro Live device which is equivalent to the RGB-D sensor present in Microsoft Kinect. We multiply this threshold by ten, in order to address the presence of outliers and multiple clusters, and utilize $\varphi = 200\ mm$.

A.5 Segmentation of a Color Image (ics)

The appropriateness of an Euclidean model for color homogeneity has been disputed in the literature. Klinker et al. [121] presented a new approach to measuring highlights from an arbitrary point of a dielectric object in 1988. Then, in 1990, they applied their theory to color image understanding [122]. About a decade later, Cheng et al. [123] used Principal Component Analysis (PCA) for color image processing. Then, in 2004, Nikolaev et al. [124] provided theoretical evidence for the appropriateness of linear models of color homogeneity in natural images. Later in 2008, Abadpour et al. [125] utilized a resulting cylindrical cluster model for color image segmentation for the purposes of compression and watermarking. Here, we utilize this model. In order to produce the data items, we downsample the input image at the scale of 4 and utilize $w_n \equiv 1$.

A cluster in this problem class is represented as a line in $\mathbb{R}^3$. Therefore, the corresponding cluster fitting function utilizes Fuzzy Principal Component Analysis (FPCA) [126, 127]. Moreover, $\Psi_\circ(\cdot)$ utilizes the PCA representations of a maximum of 14 homogenous 256×256 colored patches. These patches are extracted form images found on the web through searching for terms such as *Skin*, *Red*, and *Olive*.

In this problem class, we utilize a tenth of the dynamic range of the input data in order to assign a value to φ. Hence, taking into consideration the dimension of x_n, we utilize $\varphi = \frac{1}{10} 255\sqrt{3}$.

A.6 Histogram-based Segmentation of a Grayscale Image (ighe)

The problem of grayscale image multi-level thresholding defines the data items as grayscale values and models a cluster as an interval on the grayscale axis centered at the scalar ψ_c [71].

In order to produce the data, we calculate the histogram of the input image (32 bins). Here, each bin represents one data item and the number of pixels in the bin, normalized over the area of the image, produces the corresponding weight. In this problem class, $\Psi_\circ(\cdot)$ generates C uniformly distributed points in the $[0, 255]$ interval and $\Psi(\cdot)$ calculates the sample mean.

Following a similar process to the one described in Appendix A.5 for the ics problem class, we utilize $\varphi = \frac{1}{10} 255$ for this problem class.

Bibliography

[1] J. B. MacQueen, Some methods for classification and analysis of multivariate observations, in: Proceedings of 5-th Berkeley Symposium on Mathematical Statistics and Probability, Berkeley, 1967, pp. 281–297.

[2] R. Gray, Y. Linde, Vector quantizers and predictive quantizers for Gauss-Markov sources, IEEE Transactions on Communications 30 (2) (1982) 381–389.

[3] T. Kanungo, D. M. Mount, N. S. Netanyahu, C. D. Piatko, R. Silverman, A. Y. Wu, An efficient k-means clustering algorithm: Analysis and implementation, IEEE Transactions on Pattern Analysis and Machine Intelligence 24 (7) (2002) 881–892.

[4] G. H. Ball, D. J. Hall, A clustering technique for summarizing multivariate data, Behavioral Science 12 (2) (1967) 153–155.

[5] L. A. Zadeh, Fuzzy sets, Information Control 8 (1965) 338–353.

[6] M.-S. Yang, A survey of fuzzy clustering, Mathematical and Computer Modelling 18 (11) (1993) 1–16.

[7] A. Baraldi, P. Blonda, A survey of fuzzy clustering algorithms for pattern recognition. I, IEEE Transactions on Systems, Man, and Cybernetics, Part B: Cybernetics 29 (6) (1999) 778–785.

[8] A. Baraldi, P. Blonda, A survey of fuzzy clustering algorithms for pattern recognition. II, IEEE Transactions on Systems, Man, and Cybernetics, Part B: Cybernetics 29 (6) (1999) 786–801.

[9] R. Duda, P. Hart, Pattern Classification and Scene Analysis, Wiley, New York, 1973.

[10] J. C. Bezdek, Pattern Recognition with Fuzzy Objective Function Algorithms, Plenum Press, New York, 1981.

[11] J. C. Bezdek, J. M. Keller, R. Krishnapuram, N. R. Pal, Fuzzy Models and Algorithms for Pattern Recognition and Image Processing, Kluwer Academic Publishers, Boston, 1999.

[12] E. H. Ruspini, A new approach to clustering, Information & Control 15 (1) (1969) 22–32.

[13] J. C. Dunn, A fuzzy relative of the ISODATA process and its use in detecting compact well-separated clusters, Journal of Cybernetics 3 (3) (1973) 32–57.

[14] C. Borgelt, Objective functions for fuzzy clustering, in: C. Moewes, A. Nurnberger (Eds.), Computational Intelligence in Intelligent Data Analysis, Vol. 445 of Studies in Computational Intelligence, Springer Berlin Heidelberg, 2013, pp. 3–16.

[15] R. Krishnapuram, J. M. Keller, The possibilistic C-means algorithms: Insights and recommendations, IEEE Transactions on Fuzzy Systems 4 (3) (1996) 385–393.

[16] J. M. Leski, Generalized weighted conditional fuzzy clustering, IEEE Transactions on Fuzzy Systems 11 (6) (2003) 709–715.

[17] J. Yu, Q. Cheng, H. Huang, Analysis of the weighting exponent in the FCM, IEEE Transactions on Systems, Man, and Cybernetics, Part B: Cybernetics 34 (1) (2004) 634–639.

[18] M. Trivedi, J. C. Bezdek, Low-level segmentation of aerial images with fuzzy clustering, IEEE Transactions on Systems, Man, and Cybernetics 16 (4) (1986) 589–598.

[19] H. Frigui, R. Krishnapuram, A robust algorithm for automatic extraction of an unknown number of clusters from noisy data, Pattern Recognition Letters 17 (12) (1996) 1223–1232.

[20] F. Klawonn, R. Kruse, H. Timm, Fuzzy shell cluster analysis, in: G. della Riccia, H. Lenz, R. Kruse (Eds.), Learning, networks and statistics, Springer, 1997, pp. 105–120.

[21] J. C. Bezdek, A physical interpretation of fuzzy ISODATA, IEEE Transactions on Systems, Man and Cybernetics SMC-6 (5) (1976) 387–389.

[22] N. R. Pal, J. C. Bezdek, On cluster validity for the fuzzy C-means model, IEEE Transactions on Fuzzy Systems 3 (3) (1995) 370–379.

[23] K. Zhou, C. Fu, S. L. Yang, Fuzziness parameter selection in fuzzy c-means: The perspective of cluster validation, Science China Information Sciences 57 (11) (2014) 1–8.

[24] M. Lichman, UCI machine learning repository (2013).
URL http://archive.ics.uci.edu/ml

[25] J. C. Bezdek, N. R. Pal, Some new indexes of cluster validity, IEEE Transactions on Systems, Man, and Cybernetics, Part B: Cybernetics 28 (3) (1998) 301–315.

[26] I. Sledge, J. C. Bezdek, T. C. Havens, J. M. Keller, Relational generalizations of cluster validity indices, IEEE Transactions on Fuzzy Systems 18 (4) (2010) 771–786.

[27] I. Ozkan, I. Turksen, Upper and lower values for the level of fuzziness in FCM, in: P. P. Wang, D. Ruan, E. E. Kerre (Eds.), Fuzzy Logic, Vol. 215 of Studies in Fuzziness and Soft Computing, Springer Berlin Heidelberg, 2007, pp. 99–112.

[28] K.-L. Wu, Analysis of parameter selections for fuzzy c-means, Pattern Recognition 45 (1) (2012) 407–415.

[29] W. Pedrycz, H. Izakian, Cluster-centric fuzzy modeling, IEEE Transactions on Fuzzy Systems 22 (6) (2014) 1585–1597.

[30] P. J. Rousseeuw, E. Trauwaert, L. Kaufman, Fuzzy clustering with high contrast, Journal of Computational and Applied Mathematics 64 (1-2) (1995) 81–90.

[31] F. Klawonn, F. Hoppner, What is fuzzy about fuzzy clustering? Understanding and improving the concept of the fuzzifier, in: M. R. Berthold, H.-J. Lenz, E. Bradley, R. Kruse, C. Borgelt (Eds.), Advances in Intelligent Data Analysis V, Vol. 2810 of Lecture Notes in Computer Science, Springer Berlin Heidelberg, 2003, pp. 254–264.

[32] F. Klawonn, Fuzzy clustering: Insights and a new approach, Mathware and soft computing 11 (2004) 125–142.

[33] W. Pedrycz, Conditional fuzzy C-means, Pattern Recognition Letters 17 (6) (1996) 625–631.

[34] W. Pedrycz, Fuzzy set technology in knowledge discovery, Fuzzy Sets and Systems 98 (3) (1998) 279–290.

[35] W. Pedrycz, Conditional fuzzy clustering in the design of radial basis function neural networks, IEEE Transactions on Neural Networks 9 (4) (1998) 601–612.

[36] K. K. Chintalapudi, M. Kam, The credibilistic fuzzy C-means clustering algorithm, in: IEEE International Conference on Systems, Man, and Cybernetics (SMC 1998), Vol. 2, 1998, pp. 2034–2039.

[37] K. K. Chintalapudi, M. Kam, A noise-resistant fuzzy C means algorithm for clustering, in: Proceedings of IEEE World Congress on Computational Intelligence, Vol. 2, 1998, pp. 1458–1463.

[38] M.-S. Yang, K.-L. Wu, Unsupervised possibilistic clustering, Pattern Recognition 39 (1) (2006) 5–21.

[39] J. Noordam, W. van den Broek, L. Buydens, Multivariate image segmentation with cluster size insensitive fuzzy C-means, Chemometrics and Intelligent Laboratory Systems 64 (1) (2002) 65–78.

[40] R. Kruse, C. Doring, M.-J. Lesot, Fundamentals of fuzzy clustering, in: J. V. de Oliveira, W. Pedrycz (Eds.), Advances in Fuzzy Clustering and its Applications, Wiley, England, 2007, pp. 3–29.

[41] R. Yager, D. Filev, Approximate clustering via the mountain method, IEEE Transactions on Systems, Man and Cybernetics 24 (8) (1994) 1279–1284.

[42] G. Beni, X. Liu, A least biased fuzzy clustering method, IEEE Transactions on Pattern Analysis and Machine Intelligence 16 (9) (1994) 954–960.

[43] K. Rose, E. Gurewitz, G. Fox, A deterministic annealing approach to clustering, Pattern Recognition Letters 11 (9) (1990) 589–594.

[44] K. Rose, E. Gurewitz, G. Fox, Constrained clustering as an optimization method, IEEE Transactions on Pattern Analysis and Machine Intelligence 15 (8) (1993) 785–794.

[45] J. M. Leski, Fuzzy c-varieties/elliptotypes clustering in reproducing kernel Hilbert space, Fuzzy Sets and Systems 141 (2) (2004) 259–280.

[46] D.-M. Tsai, C.-C. Lin, Fuzzy C-means based clustering for linearly and nonlinearly separable data, Pattern Recognition 44 (8) (2011) 1750–1760.

[47] K.-L. Wu, M.-S. Yang, Alternative C-means clustering algorithms, Pattern Recognition 35 (10) (2002) 2267–2278.

[48] L. Chen, C. Chen, M. Lu, A multiple-kernel fuzzy C-means algorithm for image segmentation, IEEE Transactions on Systems, Man, and Cybernetics, Part B: Cybernetics 41 (5) (2011) 1263–1274.

[49] S. Chen, D. Zhang, Robust image segmentation using FCM with spatial constraints based on new kernel-induced distance measure, IEEE Transactions on Systems, Man, and Cybernetics, Part B: Cybernetics 34 (4) (2004) 1907–1916.

[50] K. Honda, N. Sugiura, H. Ichihashi, Fuzzy PCA-guided robust k-means clustering, IEEE Transactions on Fuzzy Systems 18 (1) (2010) 67–79.

[51] H. Zha, C. Ding, M. Gu, X. He, H. Simon, Spectral relaxation for K-means clustering, in: Proceedings of Advances in Neural Information Processing Systems, 2002, pp. 1057–1064.

[52] R. J. Hathaway, J. W. Davenport, J. C. Bezdek, Relational duals of the C-means clustering algorithms, Pattern Recognition 22 (2) (1989) 205–212.

[53] R. J. Hathaway, J. C. Bezdek, NERF C-means: Non-Euclidean relational fuzzy clustering, Pattern Recognition 27 (3) (1994) 429–437.

[54] S. Nascimento, B. Mirkin, F. Moura-Pires, Multiple prototype model for fuzzy clustering, in: D. J. Hand, J. N. Kok, M. R. Berthold (Eds.), Advances in Intelligent Data Analysis, Vol. 1642 of Lecture Notes in Computer Science, Springer Berlin Heidelberg, 1999, pp. 269–279.

[55] K. Jajuga, L_1-norm based fuzzy clustering, Fuzzy Sets and Systems 39 (1) (1991) 43–50.

[56] L. Bobrowski, J. C. Bezdek, C-means clustering with the ℓ_1 and ℓ_∞ norms, IEEE Transactions on Systems, Man, and Cybernetics 21 (3) (1991) 545–554.

[57] R. J. Hathaway, J. C. Bezdek, Optimization of clustering criteria by reformulation, IEEE Transactions on Fuzzy Systems 3 (1995) 241–246.

[58] R. J. Hathaway, J. C. Bezdek, Y. Hu, Generalized fuzzy C-means clustering strategies using L_p norm distances, IEEE Transactions on Fuzzy Systems 8 (5) (2000) 576–582.

[59] N. B. Karayiannisa, M. M. Randolph-Gips, Non-Euclidean C-means clustering algorithms, Intelligent Data Analysis 7 (2003) 405–425.

[60] D. E. Gustafson, W. C. Kessel, Fuzzy clustering with a fuzzy covariance matrix, in: IEEE Conference on Decision and Control including the 17th Symposium on Adaptive Processes, Vol. 17, San Diego, CA, 1979, pp. 761–766.

[61] I. Gath, A. Geva, Unsupervised optimal fuzzy clustering, IEEE Transaction on Pattern Analysis Machine Intelligence 11 (7) (1989) 773–781.

[62] H. Frigui, R. Krishnapuram, A comparison of fuzzy shell-clustering methods for the detection of ellipses, IEEE Transactions on Fuzzy Systems 4 (2) (1996) 193–199.

[63] R. Krishnapuram, H. Frigui, O. Nasraoui, Fuzzy and possibilistic shell clustering algorithms and their application to boundary detection and surface approximation - Parts I & II, IEEE Transaction on Fuzzy Systems 3 (1) (1995) 29–60.

[64] R. N. Dave, R. Krishnapuram, Robust clustering methods: A unified view, IEEE Transactions on Fuzzy Systems 5 (2) (1997) 270–293.

[65] J. Leski, Towards a robust fuzzy clustering, Fuzzy Sets and Systems 137 (2) (2003) 215–233.

[66] P. D'Urso, L. D. Giovanni, Robust clustering of imprecise data, Chemometrics and Intelligent Laboratory Systems 136 (2014) 58–80.

[67] J. J. D. Gruijter, A. B. McBratney, A modified fuzzy K-means method for predictive classification, in: H. H. Bock (Ed.), Classification and Related Methods of Data Analysis, Elsevier, Amsterdam, The Netherlands, 1988, pp. 97–104.

[68] R. N. Dave, Characterization and detection of noise in clustering, Pattern Recognition Letters 12 (11) (1991) 657–664.

[69] Y. Ohashi, Fuzzy clustering and robust estimation, Presented at the 9th SAS Users Group International (SUGI) Meeting at Hollywood Beach, Florida. (1984).

[70] R. N. Dave, Robust fuzzy clustering algorithms, in: Second IEEE International Conference on Fuzzy Systems, Vol. 2, 1993, pp. 1281–1286.

[71] J. M. Jolion, P. Meer, S. Bataouche, Robust clustering with applications in computer vision, IEEE Transactions on Pattern Analysis and Machine Intelligence 13 (8) (1991) 791–802.

[72] S. Zhuang, T. Wang, P. Zhang, A highly robust estimator through partially likelihood function modeling and its application in computer vision, IEEE Transactions on Pattern Analysis and Machine Intelligence 14 (1) (1992) 19–35.

[73] R. Krishnapuram, J. M. Keller, A possibilistic approach to clustering, IEEE Transactions on Fuzzy Systems 1 (2) (1993) 98–110.

[74] M. Barni, V. Cappellini, A. Mecocci, Comments on "A possibilistic approach to clustering", IEEE Transactions on Fuzzy Systems 4 (3) (1996) 393–396.

[75] H. Timm, C. Borgelt, C. Doring, R. Kruse, An extension to possibilistic fuzzy cluster analysis, Fuzzy Sets and Systems 147 (1) (2004) 3–16.

[76] R. Dave, S. Sen, On generalising the noise clustering algorithms, in: Proceedings of the 7th IFSA World Congress (IFSA 1997), 1997, pp. 205–210.

[77] N. R. Pal, K. Pal, J. C. Bezdek, A mixed c-means clustering model, in: Proceedings of the Sixth IEEE International Conference on Fuzzy Systems, Vol. 1, 1997, pp. 11–21.

[78] N. R. Pal, K. Pal, J. M. Keller, J. C. Bezdek, A new hybrid C-means clustering model, in: Proceedings of the 2004 IEEE International Conference on Fuzzy Systems, Vol. 1, 2004, pp. 179–184.

[79] X.-Y. Wang, J. M. Garibaldi, Simulated annealing fuzzy clustering in cancer diagnosis, Informatica 29 (1) (2005) 61–70.

[80] A. Keller, Fuzzy clustering with outliers, in: Proceesings of the 19th International Conference of the North American Fuzzy Information Processing Society (NAFIPS 2000), 2000, pp. 143–147.

[81] D.-Q. Zhang, S.-C. Chen, A comment on "Alternative C-means clustering algorithms", Pattern Recognition 37 (2) (2004) 173–174.

[82] R. Krishnapuram, C.-P. Freg, Fitting an unknown number of lines and planes to image data through compatible cluster merging, Pattern Recognition 25 (4) (1992) 385–400.

[83] R. Krishnapuram, O. Nasraoui, H. Frigui, The fuzzy C-spherical shells algorithm: a new approach, IEEE Transactions on Neural Networks 3 (5) (1992) 663–671.

[84] R. N. Dave, T. Fu, Robust shape detection using fuzzy clustering: Practical applications, Fuzzy Sets and Systems 65 (2-3) (1994) 161–185.

[85] H. Frigui, R. Krishnapuram, Clustering by competitive agglomeration, Pattern Recognition 30 (7) (1997) 1109–1119.

[86] P. D'Urso, Fuzzy clustering of fuzzy data, in: J. V. de Oliveira, W. Pedrycz (Eds.), Advances in Fuzzy Clustering and its Applications, Wiley, England, 2007, pp. 155–192.

[87] A. Abadpour, A. S. Alfa, J. Diamond, Video-on-demand network design and maintenance using fuzzy optimization, IEEE Transactions on Systems, Man, and Cybernetics, Part B: Cybernetics 38 (2) (2008) 404–420.

[88] L. Szilagyi, Z. Benyo, S. Szilagyi, H. S. Adam, MR brain image segmentation using an enhanced fuzzy C-means algorithm, in: Proceedings of the 25th Annual International Conference of the IEEE Engineering in Medicine and Biology Society (EMBS 2003), Vol. 1, 2003, pp. 724–726.

[89] W. Cai, S. Chen, D. Zhang, Fast and robust fuzzy C-means clustering algorithms incorporating local information for image segmentation, Pattern Recognition 40 (3) (2007) 825–838.

[90] R. J. Hathaway, Y. Hu, Density-weighted fuzzy C-means clustering, IEEE Transactions on Fuzzy Systems 17 (1) (2009) 243–252.

[91] Y. Yang, Image segmentation based on fuzzy clustering with neighborhood information, Optica Applicata 39 (1) (2009) 135–147.

[92] R. Nock, F. Nielsen, On weighting clustering, IEEE Transactions on Pattern Analysis and Machine Intelligence 28 (8) (2006) 1223–1235.

[93] J.-L. Chen, J.-H. Wang, A new robust clustering algorithm-density-weighted fuzzy C-means, in: Proceedings of IEEE International Conference on Systems, Man, and Cybernetics (SMC 1999), Vol. 3, 1999, pp. 90–94.

[94] A. H. Hadjahmadi, M. M. Homayounpour, S. M. Ahadi, Bilateral weighted fuzzy C-means clustering, Iranian Journal of Electrical & Electronic Engineering 8 (2012) 108–121.

[95] A. M. Bensaid, L. O. Hall, J. C. Bezdek, L. P. Clarke, Partially supervised clustering for image segmentation, Pattern Recognition 29 (5) (1996) 859–871.

[96] C.-H. Li, W.-C. Huang, B.-C. Kuo, C.-C. Hung, A novel fuzzy weighted C-means method for image classification, International Journal of Fuzzy Systems 10 (3) (2008) 168–173.

[97] C.-C. Hung, S. Kulkarni, B.-C. Kuo, A new weighted fuzzy C-means clustering algorithm for remotely sensed image classification, IEEE Journal of Selected Topics in Signal Processing 5 (3) (2011) 543–553.

[98] A. Abadpour, Rederivation of the fuzzypossibilistic clustering objective function through Bayesian inference, Fuzzy Sets and Systems 305 (2016) 29–53.

[99] A. Abadpour, A sequential bayesian alternative to the classical parallel fuzzy clustering model, Information Sciences 318 (2015) 28–47.

[100] P. W. Holland, R. E. Welsch, Robust regression using iteratively reweighted least squares, Communication Statistics - Theory and Methods A6 (9) (1977) 813–827.

[101] C. J. Friedrich, Alfred Weber's Theory of the Location of Industries, Chicago University Press, Chicago, 1929, translated from the original title "Uber den Standort der Industrien" by Alfred Weber.

[102] G. Wesolowski, The Weber problem: History and perspective, Location Science 1 (1993) 5–23.

[103] E. Weiszfeld, Sur le point pour lequel la somme des distances de n points donnes est minimum, Tohoku Mathematical Journal 43 (1937) 355–386.

[104] Z. Drezner, A note on accelerating the Weiszfeld procedure, Location Science 3 (1995) 275–279.

[105] M. R. Osborne, Finite Algorithms in Optimization and Data Analysis, John Wiley, New York, 1985.

[106] A. N. Tikhonov, V. Y. Arsenin, Solutions of ill posed problems, Mathematics of Computation 32 (144) (1977) 1320–1322.

[107] K. Levenberg, A method for the solution of certain non-linear problems in least squares, Quarterly Journal of Applied Mathmatics II (2) (1944) 164–168.

[108] D. W. Marquardt, An algorithm for least-squares estimation of nonlinear parameters, Journal of the Society for Industrial and Applied Mathematics 11 (2) (1963) 431–441.

[109] J. J. More, The Levenberg-Marquardt algorithm: Implementation and theory, in: G. Watson (Ed.), Numerical Analysis, Vol. 630 of Lecture Notes in Mathematics, Springer Berlin Heidelberg, 1978, pp. 105–116.

[110] A. Beck, S. Sabach, Weiszfelds method: Old and new results, Journal of Optimization Theory and Applications (2014) 1–40.

[111] P. J. Huber, E. Ronchetti, Robust Statistics, Wiley, New York, 2009.

[112] A. E. Beaton, J. W. Tukey, The fitting of power series, meaning polynomials, illustrated on band-spectroscopic data, Technometrics 16 (1974) 147–185.

[113] F. R. Hampel, E. M. Ponchotti, P. J. Rousseeuw, W. A. Stahel, Robust Statistics: The Approach based on Influence Functions, Wiley, New York, 2005.

[114] A. Blake, A. Zisserman, Visual Reconstruction, MIT Press, Cambridge, USA, 1987.

[115] I. Kasa, A circle fitting procedure and its error analysis, IEEE Transactions on Instrumentation and Measurement IM-25 (1) (1976) 8–14.

[116] C. Rusu, M. Tico, P. Kuosmanen, E. J. Delp, Classical geometrical approach to circle fitting– review and new developments, Journal of Electronic Imaging 12 (1) (2003) 179–193.

[117] W. Gander, G. H. Golub, R. Strebel, Least-squares fitting of circles and ellipses, BIT Numerical Mathematics 34 (4) (1994) 558–578.

[118] D. Umbach, K. N. Jones, A few methods for fitting circles to data, IEEE Transactions on Instrumentation and Measurement 52 (6) (2003) 1881–1885.

[119] R. Maier, Out-of-core bundle adjustment for 3D workpiece reconstruction, Master's thesis, Technische Universitat Munchen, Germany (2013).

[120] M. A. Fischler, R. C. Bolles, Random Sample Consensus: A paradigm for model fitting with applications to image analysis and automated cartography, Communications of the ACM 24 (6) (1981) 381–395.

[121] G. J. Klinker, S. A. Shafer, T. Kanade, The measurement of highlights in color images, International Journal of Computer Vision 2 (1988) 7–32.

[122] G. J. Klinker, S. A. Shafer, T. Kanade, A physical approach to color image understanding, International Journal of Computer Vision 4 (1990) 7–38.

[123] S.-C. Cheng, S.-C. Hsia, Fast algorithm's for color image processing by principal component analysis, Journal of Visual Communication and Image Representation 14 (2003) 184–203.

[124] D. O. Nikolaev, P. O. Nikolayev, Linear color segmentation and its implementation, Computer Vision and Image Understanding 94 (2004) 115–139.

[125] A. Abadpour, S. Kasaei, Color PCA eigenimages and their application to compression and watermarking, IEE Image & Vision Computing 26 (7) (2008) 878–890.

[126] Y. Yabuuchi, J. Watada, Fuzzy principal component analysis and its application, Biomedical Fuzzy and Human Sciences 3 (1997) 83–92.

[127] T. Cundari, C. Sarbu, H. F. Pop, Robust fuzzy principal component analysis (FPCA). A comparative study concerning interaction of carbon-hydrogen bonds with molybdenum-oxo bonds, Journal of Chemical Information and Computer Sciences 42 (6) (2002) 1363–1369.

www.ingramcontent.com/pod-product-compliance
Lightning Source LLC
Chambersburg PA
CBHW061445180526
45170CB00004B/1561